Spectroscopic methods in organic chemistry

Second edition

Consulting Editor
P. Sykes, M.Sc., Ph.D.,
Fellow of Christ's College
University of Cambridge

Other titles in the European Chemistry Series

Avery: The Quantum Theory of Atoms, Molecules and Photons
Banwell: Fundamentals of Molecular Spectroscopy, Second Edition
Barnard : Theoretical Basis of Inorganic Chemistry
Barnard & Chayen: Modern Methods of Chemical Analysis
Benson: Mechanisms of Inorganic Reactions in Solution
Brennan & Tipper: A Laboratory Manual of Experiments in Physical Chemistry
Bu'Lock: The Biosynthesis of Natural Products
Chapman: Introduction to Lipids
Cox & Kemp: Introductory Photochemistry
Garratt: Aromaticity
Gimblett: Introduction to the Kinetics of Chemical Chain Reactions
Hallas: Organic Stereochemistry
Kemp: Practical Organic Chemistry
Kemp: Qualitative Organic Analysis
Krebs: Fundamentals of Inorganic Crystal Chemistry
Ross: Inorganic Infrared and Raman Spectra
Sutton: Electronic Spectra of Transition Metal Complexes
Williams: Introduction to the Chemistry of Enzyme Action
Williams & Fleming: Spectroscopic Problems in Organic Chemistry
Williams & Howe : Principles of Organic Mass Spectrometry

Spectroscopic methods in organic chemistry

Second edition

Dudley H. Williams, M.A., Ph.D., Sc.D.
(Fellow of Churchill College, Cambridge)

Ian Fleming, M.A., Ph.D.
(Fellow of Pembroke College, Cambridge)

London · New York · St Louis · San Francisco · Düsseldorf ·
Johannesburg · Kuala Lumpur · Mexico · Montreal · New Delhi ·
Panama · Paris · São Paulo · Singapore · Sydney · Toronto

Published by

McGraw-Hill Book Company (UK) Limited
MAIDENHEAD . BERKSHIRE . ENGLAND

07 084023 7

PRINTED AND BOUND IN GREAT BRITAIN

Preface

We have written this book as a guide to the interpretation of the ultraviolet, infrared, nuclear magnetic resonance, and mass spectra of organic compounds; it is intended both as a textbook suitable for a first course in the subject, and as a handbook for practising organic chemists.

Spectroscopic methods are now used at some point in the solution of almost all problems in organic chemistry. Two of these methods, ultraviolet spectroscopy and infrared spectroscopy, rely on the selective absorption of electromagnetic radiation by organic molecules. The former method is used to detect conjugated systems, because the promotion of electrons from the ground state to the excited state of such systems gives rise to absorption in this region. The latter is used to detect and identify the vibrations of molecules, and especially the characteristic vibrations of the double and triple bonds present in many functional groups. The third method, nuclear magnetic resonance spectroscopy, uses a longer wavelength of the electromagnetic spectrum to detect changes in the alignment of nuclear magnets in strong magnetic fields. Absorption is observed from such nuclei as ^{1}H, ^{13}C, ^{15}N, ^{19}F, and ^{31}P; and the precise frequency of absorption is a very sensitive measure of the magnetic, and hence the chemical, environment of such nuclei. Moreover, the number and disposition of neighbouring magnetic nuclei influence the appearance of that absorption in a well-defined way. The result, particularly with the ubiquitous hydrogen nuclei, is a very considerable gain in information about the arrangement of functional groups and hydrocarbon residues in a molecule. The fourth method, mass spectrometry, measures the mass-to-charge ratio of organic ions created by electron bombardment. Structural information comes from the moderately predictable fragmentation organic molecules undergo; the masses of the fragment ions can often be related to likely structures.

Other regions of the electromagnetic spectrum are often used to determine the structure of organic molecules. X-ray diffraction can be used to pinpoint centres of high electron density (that is, the atoms). Microwave absorption is used to measure molecular rotations. Electron spin resonance, also using radio frequency signals, detects unpaired electrons, and can be used to measure the distribution of electron density in radicals. Optical rotatory dispersion and circular dichroism, using visible and ultraviolet light, measure the change in rotatory power of molecules, as the wavelength of the polarized light is changed; such measurements can often be related to the absolute configuration of molecules. Other physical methods, such as the measurement of pKs, reaction rates and dipole moments, are also used by organic chemists for structure determination. But all of these methods are more specialized than the four spectroscopic methods described in this book: these four methods are so regularly used that all organic chemists need to know about them.

We have kept discussion of the theoretical background to a minimum, since correlations between spectra and structure can successfully be made without detailed theoretical knowledge; this aspect of the subject has, moreover, been covered in many books, including one in this series (C. N. Banwell, *Fundamentals of Molecular Spectroscopy*). We have instead discussed in each chapter the kind of information given by each of the four spectroscopic methods, and we have described how to read each kind of spectrum to get that information out of it. We have included in each chapter tables of values: UV maxima, IR frequencies, NMR chemical shifts and coupling constants, and common mass fragments found in mass spectra—all of which are regularly needed for the day-to-day interpretation of spectra.

Finally, in chapter 5, we give practical examples of the way in which these four methods can be brought together to solve some moderately difficult structural problems. In a companion volume (D. H. Williams and I. Fleming, *Spectroscopic Problems in Organic Chemistry*), we have prepared a graded series of spectra, including 24 of the simplest kind, which students can use to practise what they have learned from this book; together the two books make up a course in spectroscopic methods. The 12 sets of spectra at the end are offered for amusement. They are intended to supplement an organized course which the student is attending. Throughout the book we have stressed the application of spectroscopic methods to structure determination, though the application to other problems is limited only by the ingenuity of the researcher and analyst.

In preparing a second edition, we have dispensed with what was the first chapter in the first edition, an essay on physical methods; such an essay is redundant now that physical methods are so well accepted in organic chemistry. The text of the chapters on UV and IR is changed only in detail, but we have retaken all of the IR spectra on a linear-in-frequency machine, in line with most current practice. We have made several additions to the chapter on NMR and largely rewritten the chapter on mass spectroscopy. These two methods are the newest, and they have advanced substantially since our first edition.

Cambridge   IAN FLEMING
                                         DUDLEY H. WILLIAMS

Contents

1. Ultraviolet and Visible Spectra

1–1. Introduction

The visible and ultraviolet spectra of organic compounds are associated with transitions between electronic energy levels. The transitions are generally between a bonding or lone-pair orbital and an unfilled non-bonding or anti-bonding orbital. The wavelength of the absorption is then a measure of the separation of the energy levels of the orbitals concerned. The highest energy separation is found when electrons in σ-bonds are excited, giving rise to absorption in the 120 to 200 nm (1 nm = 10^{-7} cm. = 10 Å = 1 mμ) range. This range, known as the vacuum ultraviolet, since air must be excluded from the instrument, is both difficult to measure and relatively uninformative. Above 200 nm, however, excitation of electrons from p- and d-orbitals and π-orbitals, and, particularly, π-conjugated systems, gives rise to readily measured and informative spectra.

1–2. The Energy of Electronic Excitation

The energy is related to wavelength by equation 1–1.

$$E \text{ (kcals/mole)} = \frac{28 \cdot 6 \times 1000}{\lambda \text{ (nm)}} \qquad (1\text{–}1)$$

Thus 286 nm, for example, is equivalent to 100 kcals ($\approx$ 420 kJ)—enough energy, incidentally, to initiate many interesting reactions; compounds should not, therefore, be left in the ultraviolet beam any longer than is necessary.

1–3. The Absorption Laws

Two empirical laws have been formulated about the absorption intensity. *Lambert's law* states that the fraction of the incident light absorbed is independent of the intensity of the source. *Beer's law* states that the absorption is proportional to the number of absorbing molecules. From these laws, the remaining variables give the equation 1–2.

$$\log_{10} \frac{I_0}{I} = \varepsilon . l . c \qquad (1\text{–}2)$$

I_0 and I are the intensities of the incident and transmitted light respectively, l is the path length of the absorbing solution in centimetres, and c is the concentration in moles/litre. $\log_{10}(I_0/I)$ is called the absorbance or optical density; ε is known as the molar extinction coefficient and has units of 1000 cm.^{2}/mole but the units are, by convention, never expressed.

1–4. Measurement of the Spectrum

The ultraviolet or visible spectrum is usually taken on a very dilute solution. An appropriate quantity of the compound (often about 1 mg. when the compound has a molecular weight of 100 to 200) is weighed accurately, dissolved in the solvent of choice (see below) and made up to, for instance, 100 ml. A portion of this is transferred to a silica cell. The cell is so made that the beam of light passes through a 1 cm. thickness (the value l in equation 1–2) of solution. A matched cell containing pure solvent is also prepared, and each cell is placed in the appropriate place in the spectrometer. This is so arranged that two equal beams of ultraviolet or visible light are passed, one through the solution of the sample, one through the pure solvent. The intensities of the transmitted beams are then compared over the whole wavelength range of the instrument. In most spectrometers there are two sources, one of 'white' ultraviolet and one of white visible light, which have to be changed when a complete scan is required. Usually either the visible or ultraviolet alone is sufficient for the purpose in hand. The spectrum is plotted automatically on most machines as a $\log_{10}(I_0/I)$ ordinate and λ abscissa. For publication and comparisons these are often converted to an ε versus λ or log ε versus λ plot. The unit of λ is almost always nm. Strictly speaking the intensity of a transition is better measured by the area under the absorption peak (when plotted as ε versus frequency) than by the intensity of the maximum of the peak. For several reasons, most particularly convenience and the difficulty of dealing with overlapping bands, the latter procedure is adopted in everyday use. Spectra are quoted, therefore, in terms of λ_{max}, the wavelength of the absorption peak, and ε_{max}, the intensity of the absorption peak as defined by equation 1–2.

1–5. Vibrational Fine Structure

The excitation of electrons is accompanied by changes in the vibrational and rotational quantum numbers so that what would otherwise be an absorption *line* becomes a broad peak containing vibrational and rotational fine structure. Due to interactions of solute with solvent molecules this is usually blurred out, and a smooth curve is observed. In the vapour phase, in non-polar solvents, and with certain peaks (e.g., benzene with the 260 nm band), the vibrational fine structure is sometimes observed.

1–6. Choice of Solvent

The solvent most commonly used is 95 per cent ethanol (commercial absolute ethanol contains residual benzene which absorbs in the ultraviolet). It is cheap, a good solvent and transparent down

Table 1–1

Some Solvents used in Ultraviolet Spectroscopy

Solvent	Minimum wavelength for 1 cm. cell, nm
Acetonitrile	190
Water	191
Cyclohexane	195
Hexane	201
Methanol	203
Ethanol	204
Ether	215
Methylene dichloride	220
Chloroform	237
Carbon tetrachloride	257

to about 210 nm. Fine structure, if desired, may be revealed by using cyclohexane or other hydrocarbon solvents which, being less polar, have least interaction with the absorbing molecules. Table 1–1 gives a list of common solvents and the minimum wavelength from which they may be used in 1 cm. cells.

The effect of solvent polarity on the position of maxima is discussed in section 1–9.

1–7. Selection Rules and Intensity

The irradiation of organic compounds may or may not give rise to excitation of electrons from one orbital (usually a lone-pair or bonding orbital) to another orbital (usually a non-bonding or antibonding orbital). It can be shown that:

$$\varepsilon = 0.87 \times 10^{20} \, P.\mathbf{a} \qquad (1-3)$$

where P is called the transition probability (with values from 0 to 1) and $\mathbf{a}$ is the target area of the absorbing system; the absorbing system is usually called a chromophore. With common chromophores of the order of 10 Å long, a transition of unit probability will have an ε value of 10^5. This is close to the highest observed values, though—with unusually long chromophores— values in excess of this have been measured. In practice, a chromophore giving rise to absorption by a fully allowed transition will have ε values greater than about 10,000, while those with low transition probabilities will have ε values below 1000. The important point is that, in general, *the longer a particular kind of chromophore, the more intense the absorption.*

There are many factors which affect the transition probability of any particular transition. In the first place there are rules about which transitions are allowed and which are forbidden. These are complicated because they are a function of the symmetry and multiplicity both of the ground state and excited state orbitals concerned. The spectra of chromophores, with ε_{max} less than about 10,000, are the result of 'forbidden' transitions. Two very important and 'forbidden' transitions are observed: (*a*) the n→π* band near 300 nm of ketones, with ε values of the order of 10 to 100, and (*b*) the benzene 260 nm band and its equivalent in more complicated systems, with ε values from 100 upwards. Both occur because the symmetry which makes absorption strictly forbidden is broken up by molecular vibrations and—in the latter case—by substitution.

Both types are discussed further under the sections on ketones and aromatic systems.

In this and the following discussions a very simplified theoretical picture is given; there is considerable danger in being satisfied with so little in so well developed a subject. The books by Jaffé and Orchin and by Murrell, listed in the bibliography, give excellent accounts of the state of the art.

1–8. Chromophores

The word chromophore is used to describe the system containing the electrons responsible for the absorption in question. Most of the simple unconjugated chromophores described in Table 1–2 below give rise to such high-energy, and therefore such short-wavelength absorption, that they are of little use.

Table 1–2

The Absorption of Simple Unconjugated Chromophores

Chromophore	Transition notation†	λ_{max} in nm
σ-Bonded electrons		
$>$C—C$<$ and $>$C—H	$\sigma \rightarrow \sigma^*$	~150
Lone-pair electrons		
—Ö—	$n \rightarrow \sigma^*$	~185
—N̈$<$	$n \rightarrow \sigma^*$	~195
—S̈—	$n \rightarrow \sigma^*$	~195
$>$C=Ö	$n \rightarrow \pi^*$	~300
$>$C=Ö	$n \rightarrow \sigma^*$	~190
π-bonded electrons		
$>$C=C$<$ (isolated)	$\pi \rightarrow \pi^*$	~190

† There are many other notations used

One of the few useful simple unconjugated chromophores is the very weak forbidden $n \rightarrow \pi^*$ transition of ketones mentioned earlier which appears in the 300 nm region and is of particular importance in connection with optical rotatory dispersion. This band is due to the excitation of one of the lone pair of electrons (designated n) on the oxygen atom to the lowest anti-bonding orbital (designated π^*)

of the carbonyl group. It is discussed further in the sections on solvent effects and on ketones.

The important chromophores are those in which conjugation is present. An isolated double bond or lone pair of electrons gives rise to a strong absorption maximum at about 190 nm, corresponding to the transition x in Fig. 1–1, at too short a wavelength for convenient measurement. When the molecular orbitals of two isolated double bonds are brought into conjugation, the energy level of the highest occupied orbital is raised and that of the lowest unoccupied anti-bonding orbital lowered (Fig. 1–1).

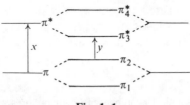

Fig. 1–1

The $\pi \rightarrow \pi^*$ transition, which is occasioned by absorption, is now associated with the smaller value y. This transition appears in the spectrum of butadiene as a strong, easily detected, and easily measured maximum at 217 nm. The same principle governs the energy levels when unlike chromophores, e.g., those of an $\alpha\beta$-unsaturated ketone, are brought together. For instance, methyl vinyl ketone has an absorption maximum at 225 nm, while neither a carbonyl group nor an isolated double bond has a strong maximum above 200 nm.

When more than two π-bonding orbitals overlap, that is when the chromophore is a longer conjugated system, the separation of the energy levels is further reduced, and absorption occurs at longer wavelength. A long conjugated polyene like carotene absorbs, quite obviously since it is coloured, in the visible. The most important point to be made is that, in general, *the longer the conjugated system, the longer the wavelength of the absorption maximum*.

The rules and correlations possible with the spectra of conjugated dienes, $\alpha\beta$-unsaturated ketones, and some substituted benzene ring compounds are given in sections 1–13, 1–16 and 1–20. With complicated chromophores, predictions become more difficult. The usual procedure, when one is confronted with the ultraviolet spectrum of an unknown substance, is to compare the spectrum, in its general shape and in the intensity and position of its peaks, with the spectra of reasonable model compounds. These models are chosen to possess as nearly as possible the same chromophore as that suspected for the unknown.

1–9. Solvent Effects

(i) $\pi \rightarrow \pi^*$. The Frank–Condon principle states that during the electronic transition atoms do not move. Electrons, however, including those of the solvent molecules, may reorganize. Most transitions result in an excited state more polar than the ground state;† the dipole–dipole interactions with solvent molecules will, therefore, lower the energy of the excited state more than that of the ground state. Thus it is usually observed that ethanol solutions give longer wavelength maxima than do hexane solutions. In other words, there is a small red-shift of the order of 10 to 20 nm in going from hexane as solvent to ethanol.

(ii) $n \rightarrow \pi^*$. The weak transition of the oxygen lone pair in ketones—the $n \rightarrow \pi^*$ transition—shows a solvent effect in the opposite direction. The solvent effect is now due to the lesser extent to which solvents can hydrogen bond to the carbonyl group in the excited state. In hexane solution, for example, the absorption maximum of acetone is at 279 nm ($\varepsilon = 15$), whereas in aqueous solution the maximum is at 264·5 nm. The shift in this direction is known as a blue shift.

1–10. Searching for a Chromophore

There is no easy rule or set procedure for identifying a chromophore—too many factors affect the spectrum and the range of structures which can be found is too great. The examination of a spectrum with particular regard for the following points is the first step to be taken.

---

† This transition is commonly visualized in valence bond terms with the ground state represented without charge separation and the excited state as the dipolar species.

Such representations are over-simplified, and misleading: the dipolar structure is not *the* structure of the excited state, rather it is a more important contributor to the excited state than to the ground state. Since the valence bond technique is less exact and less revealing in this field than is the molecular orbital theory, the latter should be used on all occasions.

(*i*) *The complexity and the extent to which the spectrum encroaches on the visible region.* A spectrum with many bands stretching into the visible shows the presence of a long conjugated or a polycyclic aromatic chromophore. A compound giving a spectrum with only one band (or only a few bands) below about 300 nm, probably contains only two or three conjugated units.

(*ii*) *The intensity of the bands, particularly the principal maximum and the longest wavelength maximum.* This observation can be very informative. Simple conjugated chromophores such as dienes and $\alpha\beta$-unsaturated ketones have ε values of 10,000 to 20,000. The longer simple conjugated systems have principal maxima (usually also the longest wavelength maxima) with correspondingly higher ε values. Very low intensity absorption bands in the 270 to 350 nm region, on the other hand, with ε values of 10 to 100, are the result of the $n \rightarrow \pi^*$ transition of ketones. In between these extremes, the existence of absorption bands with ε values of 1000 to 10,000 almost always shows the presence of an aromatic system. Many unsubstituted aromatic systems show bands with intensities of this order of magnitude, the absorption being the result of a transition with a low transition probability, low because of the symmetry of the ground and excited states. When the aromatic nucleus is substituted with groups which can extend the chromophore, strong bands with ε values above 10,000 appear, but bands with ε values below 10,000 are often still present.

Having made these observations, one should search for a model system which contains the chromophore and therefore gives a similar spectrum to that which is being examined. This may be difficult in rare cases; but so many spectra are now known, and the changes caused by substitution so well documented, that the task can be a simple one. The first tool which an organic chemist requires is a general knowledge of the simple chromophores and the changes which structural variations make in the absorption pattern. Sections 1–13 to 1–26 give a very brief account of these topics. The remaining task, that of searching through the literature, is greatly facilitated by the existence of the indexes and compilations which are described in section 1–11. The usefulness of these books will be greatly increased by a general knowledge of organic chemistry on which to base a guess as to what chromophores are likely to be known and in what compounds they may be found.

The search for a chromophore is also likely to be assisted by the other physical methods described in this book. The range of structures in which a search must be made can be narrowed, for example, to aromatic compounds on the strength of infrared or NMR aromatic C—H absorptions. Similarly the presence of an $\alpha\beta$-unsaturated ketone may be inferred from the C=O stretching vibration observed in the infrared spectrum and confirmed from the ultraviolet spectrum, and the extent of alkylation deduced by a consideration of Woodward's rules (section 1–16) and by reference to the NMR spectrum. A very important stage in determining the structure of a natural product is the positive identification of the chromophore, by comparison of the spectrum with that of some known model compound.

1–11. Standard Works of Reference

In the search for a model chromophore, a number of source books are available. In addition several textbooks, larger than the single chapter of this book, are devoted to the subject and are mentioned in the bibliography at the end of this chapter.

The major collections of *data* are the following.

(*i*) H. M. Hershenson, *Ultraviolet and Visible Absorption Spectra*, Index for 1930–1954, Academic Press, New York, 1956; Index for 1955–1959 (1961).

The index gives the names of compounds which have been examined together with literature references.

(*ii*) R. A. Friedel and M. Orchin, *Ultraviolet Spectra of Aromatic Compounds*, Wiley, New York, 1951.

This is a catalogue showing the actual spectra of 579 aromatic compounds.

(*iii*) A.P.I. Research Project 44, *Ultraviolet Spectral Data,* Carnegie Institute and U.S. Bureau of Standards.

This compilation shows the actual spectra of 917 compounds (up to October 1962), mostly aromatic hydrocarbons.

(*iv*) *Organic Electronic Spectral Data*, Interscience, New York, Vols. I–VII (1960–1971).

This most valuable collection has been prepared by a complete search of the major journals from 1945 to 1965. The compounds are indexed by their empirical formulae, and absorption maxima are quoted together with literature references.

(*v*) *UV Atlas of Organic Compounds*, Butterworths, London, 1965.

A collection of the spectra of nearly 1000 compounds elaborately cross indexed by chromophoric groups.

(*vi*) *Sadtler Standard Spectra* (Ultraviolet), Heyden, London, 1970.
A collection of the spectra of 15,000 compounds.

1–12. Definitions.

The following words and symbols are commonly used.

Red shift or *bathochromic effect*. A shift of an absorption maximum towards longer wavelength. It may be produced by a change of medium, or by the presence of an auxochrome.

Auxochrome. A substituent on a chromophore which leads to a red shift. For example, the conjugation of the lone pair on the nitrogen atom of an enamine has shifted the absorption maximum from the isolated double bond value of 190 nm to about 230 nm.· The nitrogen substituent is the auxochrome. An auxochrome, then, extends a chromophore to give a new chromophore.

Blue shift or *hypsochromic effect*. A shift towards shorter wavelength. This may be caused by a change of medium and also by such phenomena as the removal of conjugation. For example, the conjugation of the lone pair of electrons on the nitrogen atom of aniline with the π-bond system of the benzene ring is removed on protonation. Aniline absorbs at 230 nm (ε 8600), but in acid solution the main peak is almost identical with that of benzene, being now at 203 nm (ε 7500). A blue shift has occurred.

Hypochromic effect. An effect leading to decreased absorption intensity.

Hyperchromic effect. An effect leading to increased absorption intensity.

$\lambda_{\max}$ The wavelength of an absorption maximum.

ε The extinction coefficient defined by equation 1–2.

$E_{1\,cm}^{1\%}$. Absorption ($\log_{10} (I_0/I)$) of a 1 per cent solution in a cell with a 1 cm. path length. This is used in place of ε when the molecular weight of a compound is not known, or when a mixture is being examined.

Isosbestic point. A point common to all curves produced in the spectra of a compound taken at several pH values.

1-13. Conjugated Dienes

The energy levels of butadiene have been illustrated in Fig. 1–1. The transition y gives rise to strong absorption at 217 nm (ε 21,000). Alkyl substitution extends the chromophore, in the sense that there is a small interaction between the σ-bonded electrons of the alkyl group and the π-bond system. The result is a small red shift with alkyl substitution, just as there is a red shift (though a relatively large one) in going from an isolated double bond to a conjugated diene.

Fortunately the effect of alkyl substitution, in dienes at least, is additive; and a few rules suffice to predict the position of absorption in open chain dienes and dienes in six-membered rings. Open chain dienes exist normally in the s-*trans* conformation, while homoannular dienes must be in the s-*cis* conformation. These conformations are illustrated in the part structures I (heteroannular diene) and II (homoannular diene). It is not entirely clear why, but the s-*cis* conformation leads to longer wavelength absorption than does the s-*trans* conformation. Also, due to the shorter distance between the ends of the chromophore, s-*cis* dienes give maxima of lower intensity ($\varepsilon \sim 10,000$) than the maxima of s-*trans* dienes ($\varepsilon \sim 20,000$).

I   II

The actual rules for predicting the absorption of open chain and six-membered ring dienes were first made by Woodward in 1941. Since that time they have been modified by Fieser and by Scott as a result of experience with a very large number of dienes and trienes. The modified rules are given in Table 1–3.

For example the diene I would be calculated to have a maximum at 234 nm by the following addition:

Parent value	214 nm
Three ring residues (marked a) $3 \times 5 =$	15 nm
One exocyclic double bond (the Δ^4 bond is exocyclic to ring B)	5 nm
Total	234 nm

An observed value is 235 nm ($\varepsilon = 19,000$).

Table 1–3

Rules for Diene and Triene Absorption

Value assigned to parent heteroannular or open chain diene	214 nm
Value assigned to parent homoannular diene	253 nm
Increment for	
(a) each alkyl substituent or ring residue	5 nm
(b) the exocyclic nature of any double bond	5 nm
(c) a double bond extension	30 nm
(d) auxochrome —OAcyl	0 nm
—OAlkyl	6 nm
—SAlkyl	30 nm
—Cl, —Br	5 nm
—NAlkyl$_2$	60 nm
λ_{calc} Total	

(Reprinted with permission from A. I. Scott, *Interpretation of the Ultraviolet Spectra of Natural Products*, Pergamon Press, Oxford, 1964.)

By a similar calculation, the diene II would be expected to have a maximum at 273 nm, and does actually have one at 275 nm. Though ethanol is the usual solvent, change of solvent has little effect. The actual appearance of the spectrum of a simple conjugated diene with the chromophore of I is illustrated in Fig. 1–2. The discussion above has referred to the highest intensity band and, indeed, the weaker bands are not always apparent.

There are a large number of exceptions to the rules, where special factors can operate. Distortion of the chromophore may lead to red or blue shifts, depending on the nature of the distortion.

III   IV   V

The strained molecule verbenene (III) has a maximum at 245·5 nm, whereas the usual calculation gives a value of 229 nm. The diene IV might be expected to have a maximum at 273 nm; but distortion of the chromophore, presumably out of planarity with consequent loss of conjugation, causes the maximum to be as low as 220 nm with a similar loss in intensity (ε 5500). The diene V, in which coplanarity of the diene is more likely, gives a maximum at 248 nm (ε 15,800) showing that this is so although it still does not agree with the expected value. Change of ring size in the case of simple homoannular dienes also leads to departures from the predicted value of 263 nm as follows: cyclopentadiene, 238·5 nm (ε 3400); cycloheptadiene, 248 nm (ε 7500); while cyclohexadiene is close at 256 nm (ε 8000). The lesson, an important one, is that when the ultraviolet spectrum of an unknown compound is to be compared with that of a model compound, then the choice of model must be a careful one. Allowance must be made for the likely shape of the molecule and for any unusual strain. Some general comments on the effect of steric hindrance to coplanarity are given in section 1–26.

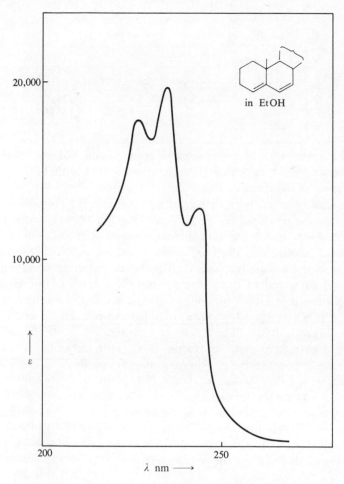

Fig. 1–2

1–14. Polyenes

As the number of double bonds in conjugation increases, the wavelength of maximum absorption encroaches on the visible region. A number of subsidiary bands also appear and the intensity increases. Table 1–4 gives examples of the longest wavelength maxima of some simple conjugated polyenes, showing these trends.

Table 1–4

Longest Wavelength Maxima of Some Simple Polyenes

	$Me(CH\overset{t}{=}CH)_n Me$		$Ph(CH\overset{t}{=}CH)_n Ph$	
n	λ_{max} nm	ε	λ_{max} nm	ε
3	274·5	30,000	358	75,000
4	310	76,500	384	86,000
5	342	122,000	403	94,000
6	380	146,500	420	113,000
7	401	—	435	135,000
8	411	—	—	—

The appearance of the spectra of some simple polyenes is illustrated in Fig. 1–3, which should be compared with the simpler spectrum of the diene in Fig. 1–2.

Several attempts both empirical and theoretical have been made to relate the principal or longest wavelength maximum with chain length. Some of the theoretical treatments have been based on the classical 'electron in the box' wave equation, in which the walls of the box are usually considered to be one average bond length beyond each end of the chromophore. This leads to the correct conclusion that increasing values of λ_{max} are found for increasing length in a conjugated polyene; quantitative predictions are, however, less satisfactory. The simple theory might indicate that as the chain length increases, the value of λ_{max} for long chains would increase proportionately; whereas in practice there is a convergence, which can be seen in Table 1–4. More sophisticated treatments, allowing for the variation in bond lengths between the double and single bonds, have been made and are described in Murrell's book. An interesting simplification is provided by the cyanine dye analogues (VI) in which overlap leads to uniform bond lengths and bond orders along the polyene chain.

$$Me_2\overset{+}{N}=CH-(CH=CH)_n-NMe_2 \longleftrightarrow Me_2N-(CH=CH)_n-CH=\overset{+}{N}Me_2$$

VI

Calculations based on the 'electron in the box' model lead to values very close to those observed: λ_{max} 309 ($n = 1$), 409 ($n = 2$) and 511 ($n = 3$) nm.

In a long chain polyene, change from *trans-* to *cis-* configuration at one or more double bonds lowers both the wavelength and the intensity of the absorption maximum.

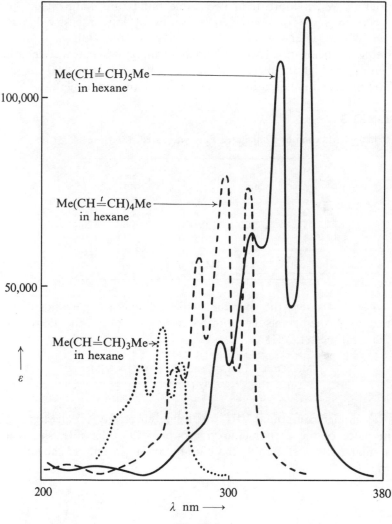

Fig. 1–3

(Replotted from Nayler and Whiting, *J. Chem. Soc.*, 1955, 3042.)

1–15. Polyeneynes and Poly-ynes

As a result of interest in natural polyeneynes and poly-ynes, the ultraviolet spectra of many such compounds are known and have been of considerable use in the elucidation of structure. The characteristic spiky appearance of the spectra has been very helpful during the screening of crude plant extracts for acetylenic compounds. When more than two triple bonds are conjugated, the spectrum shows a characteristic series of low intensity bands ($\varepsilon \sim 200$) at intervals of 2300 cm.$^{-1}$ (note the *frequency* units, frequency being directly proportional to energy whereas wavelength is not) and high intensity bands ($\varepsilon \sim 10^5$) at intervals of 2600 cm.$^{-1}$. The principal maxima in each group are shown in Table 1–5.

Table 1–5

Principal Maxima of Conjugated Poly-ynes $Me(C \equiv C)_n Me$

n	λ_{max} nm	ε	λ_{max} nm	ε
2	—	—	250	160
3	207	135,000	306	120
4	234	281,000	354	105
5	260·5	352,000	394	120
6	284	445,000	—	—

Fig. 1–4 shows the spectrum which is, like a fingerprint, diagnostic of the triyne-ene chromophore present in the dehydromatricaria ester (VII).

$$Me-C \equiv C-C \equiv C-C \equiv C-CH \overset{t}{=} CH-CO_2 Me$$

VII

A similar compound VIII, which has, effectively, an alkyl group in place of the carbomethoxyl group of the ester VII, shows a similar pattern shifted to the blue by about 15 nm, as shown on Fig. 1–4.

$$Me-C \equiv C-C \equiv C-C \equiv C-CH \overset{t}{=} CH-CH_2 CH_2 COEt$$

VIII

This is an example of the way in which an organic chemist deals with the comparison of ultraviolet spectra: most of the chromophore of VII is present in VIII, and the latter will therefore continue to show the characteristic features of the former, with a small blue shift due to the relatively small loss of conjugation.

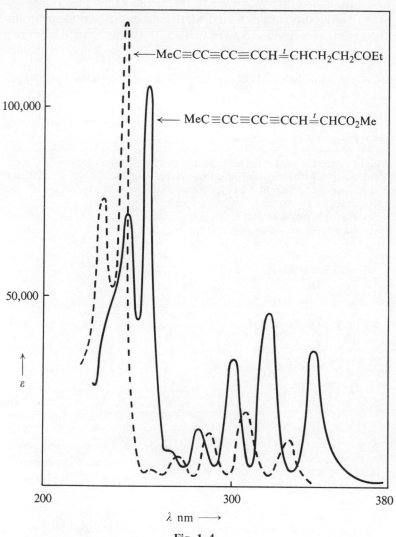

Fig. 1–4

(Replotted from Sörensen, Bruun, Holme and Sörensen, *Acta Chem. Scand.*, 1954, **8**, 28 and Bohlmann, Mannhardt and Viehe, *Chem. Ber.*, 1955, **88**, 365.)

1–16. Ketones and Aldehydes; $\pi \to \pi^*$ Transitions

Like the dienes considered in section 1–13, $\alpha\beta$-unsaturated ketones and aldehydes have been the subject of much study and their absorption, too, is susceptible to prediction by a set of rules first formulated by Woodward and modified by Fieser and by Scott. The modified rules for calculating the expected position of the absorption maximum are given in Table 1–6.

Table 1–6

Rules for $\alpha\beta$-unsaturated Ketone and Aldehyde Absorption

$$\overset{\delta}{C}=\overset{\gamma}{C}-\overset{\beta}{C}=\overset{\alpha}{C}-C=O$$

ε values are usually above 10,000 and increase with the length of the conjugated system.

Value assigned to parent $\alpha\beta$-unsaturated six-ring or acyclic ketone			215 nm
Value assigned to parent $\alpha\beta$-unsaturated five-ring ketone			202 nm
Value assigned to parent $\alpha\beta$-unsaturated aldehyde			207 nm
Increments for			
(a) a double bond extending the conjugation			30 nm
(b) each alkyl group or ring residue	α		10 nm
	β		12 nm
	γ and higher		18 nm
(c) auxochromes			
(i) —OH	α		35 nm
	β		30 nm
	δ		50 nm
(ii) —OAc	α, β, δ		6 nm
(iii) —OMe	α		35 nm
	β		30 nm
	γ		17 nm
	δ		31 nm
(iv) —SAlk	β		85 nm
(v) —Cl	α		15 nm
	β		12 nm
(vi) —Br	α		25 nm
	β		30 nm
(vii) —NR$_2$	β		95 nm
(d) the exocyclic nature of any double bond			5 nm
(e) homodiene component			39 nm

$\lambda^{\text{EtOH}}_{\text{calc}}$ Total

For $\lambda^{\text{calc}}_{\text{max}}$ in other solvents a solvent correction (Table 1–7) must be subtracted from the above value.

In this case, spectra are affected significantly by the solvent as a result of the change in polarity on excitation. A solvent correction (from Table 1–7) is subtracted from the calculated value (Table 1–6) to obtain the value expected for a solvent other than the standard solvent ethanol.

Table 1–7

Solvent Corrections for $\alpha\beta$-unsaturated Ketones

Solvent	Correction nm
Ethanol	0
Methanol	0
Dioxan	+5
Chloroform	+1
Ether	+7
Water	−8
Hexane	+11
Cyclohexane	+11

(Reprinted with permission from A. I. Scott, *Interpretation of the Ultraviolet Spectra of Natural Products*, Pergamon Press, Oxford, 1964.)

For example, mesityl oxide ($Me_2C{=}CHCOMe$) may be calculated to have λ_{max} at $(215 + 2 \times 12) = 239$ nm. The observed value is 237 nm (ε 12,600). A more complicated example, the trienone chromophore of IX, would be calculated to have a maximum at 349 nm by the following addition.

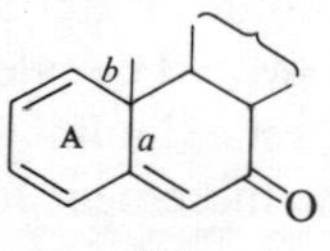

IX

Parent value	215 nm
β-substituent (marked *a*)	12 nm
ω-substituent (marked *b*)	18 nm
2 × extended conjugation	60 nm
Homoannular diene component (a special addition for this component when it is a linear part of the chromophore)	39 nm
Exocyclic double bond (the $\alpha\beta$-double bond is exocyclic to ring A)	5 nm
Total	349 nm

The observed values of λ_{max} are 230 nm (ε 18,000), 278 nm (ε 3720) and 348 nm (ε 11,000). As was the case with simple polyenes, the long chromophore present in this example gives rise to several peaks, with the longest wavelength peak in excellent agreement with prediction.

An important general principle is illustrated by the calculation for the cross-conjugated trieneone X. In this case the main chromophore is the linear dieneone portion, since the Δ^5-double bond is not in the longest conjugated system. The calculation, along the lines above, gives a value of 324 nm. The observed values are 256 nm and 327 nm. The former might be due to the Δ^5-7-one system (λ_{calc} 244 nm), but a positive identification of this sort in a complicated system is largely unjustified.

X   XI

Certain special changes in structure, as noted in the case of dienes in section 1–13, also lead to departures from the rules given above. The effect of the five-membered ring in cyclopentenones is accommodated in the rules; but when the carbonyl group is in a five-membered ring and the double bond is exocyclic to the five-membered ring, a parent value of about 215 nm holds. Another special case, verbenone, XI, would be calculated to have a maximum at 239 nm but actually has a maximum at 253 nm, an increment for strain of 14 nm, close to the increment for the corresponding diene III.

1–17. Ketones and Aldehydes; n → π* Transitions

Saturated ketones and aldehydes show a weak symmetry forbidden band, in the 275–295 nm range ($\varepsilon \sim 20$), due to excitation of an oxygen lone-pair electron to the anti-bonding π-orbital of the carbonyl group. Aldehydes and the more heavily substituted ketones absorb at the upper end of this range. Polar substituents on the α-carbon atoms raise (when axial) or lower (when equatorial) the extremes of this range. When the carbonyl group is substituted by an auxochrome as in an ester, an acid, or an amide, the π^* orbital is raised but the n level of the lone-pair left largely unaltered. The result is that the n→π^* transition of these compounds is shifted to the relatively inaccessible 200–215 nm range. The presence, therefore, of a weak band in the 275–295 nm region is positive identification of a ketone or aldehyde carbonyl group (nitro groups show a similar band and, of course, impurities must be absent). The low intensity of this transition is responsible for the ease with which the Cotton effect may be measured in studies of the optical rotatory dispersion of ketones.

$\alpha\beta$-unsaturated ketones show a slightly stronger n→π^* band or series of bands ($\varepsilon \sim 100$) in the 300–350 nm range. The precise position of these bands is not predictable from the extent of alkylation, but is a regular function of the conformation of γ-substituents, axially substituted isomers absorbing at longer wavelengths than equatorially substituted isomers.

The position and intensity of n→π^* bands are also influenced by transannular interactions (see section 1–25) and by solvent (see section 1–9 (ii)).

The n→π^* transitions of α-diketones in the diketo form give rise to two bands, one in the usual region near 290 nm ($\varepsilon \sim 30$) and a second (ε 10 to 30), which stretches into the visible in the 340–440 nm region and gives rise to the yellow colour of some of these compounds. (See also quinones in section 1–23, quinones being α-, or vinylogous α-, diketones.)

1–18. $\alpha\beta$-unsaturated Acids, Esters, Nitriles and Amides

$\alpha\beta$-unsaturated acids and esters follow a trend similar to that of the ketones but at slightly shorter wavelength. The rules for alkyl

Table 1–8

Rules for $\alpha\beta$-unsaturated Acids' and Esters' Absorption
ε values are usually above 10,000

β-monosubstituted	208 nm
$\alpha\beta$- or $\beta\beta$-disubstituted	217 nm
$\alpha\beta\beta$-trisubstituted	225 nm
Increment for	
(a) a double bond extending the conjugation	30 nm
(b) the exocyclic nature of any double bond	5 nm
(c) when the double bond is endocyclic in a five- or seven-membered ring	5 nm
λ_{calc} Total	

substitution, summarized by Nielsen, are given in Table 1–8. The change in going from acid to ester is usually not more than 2 nm.

$\alpha\beta$-unsaturated nitriles have been little studied but usually come slightly below the corresponding acids.

$\alpha\beta$-unsaturated amides have maxima lower than the corresponding acids, usually near 200 nm ($\varepsilon \sim 8000$).

$\alpha\beta$-unsaturated lactams have an additional band at 240–250 nm ($\varepsilon \sim 1000$).

1–19. The Benzene Ring

Benzene absorbs at 184 (ε 60,000), 203·5 (ε 7400) and 254 (ε 204) nm in hexane solution; it is illustrated by the solid line in Fig. 1–5. The latter band, sometimes called the B-band, shows vibrational fine structure. Although a 'forbidden' band, it owes its appearance to the loss of symmetry caused by molecular vibrations; indeed, the $0 \rightarrow 0$ transition (the transition between the ground state vibrational energy level of the electronic ground state to the ground state vibrational energy level of the electronic excited state) is not observed.

When the aromatic ring is substituted by alkyl groups, for example, or is an aza analogue such as pyridine, the symmetry is lowered; the $0 \rightarrow 0$ transition is then observed, although the spectrum is little changed otherwise. The presence of fine structure resembling that shown in Fig. 1–5 is characteristic of the simpler aromatic molecules.

When, however, the benzene ring is substituted by lone pair donating or by π-bonded systems, the chromophore is extended more usefully; unfortunately, quantitative prediction of the effects of various substituents is not always possible in the manner so successful with dienes and unsaturated ketones. Section 1–20 gives an account of some of the trends observed in compounds containing a substituted benzene ring.

1–20. Substituted Benzene Rings

Table 1–9, giving the wavelength of absorption maxima in the spectra of a range of monosubstituted benzenes, shows how, as usual, the wavelength and intensity of the absorption peaks increase with an increase in the extent of the chromophore. As more and more conjugation is added to the benzene ring, the band originally at 203·5 nm (sometimes called the K-band) effectively 'moves' to longer wavelength, and moves 'faster' than the B-band, which was

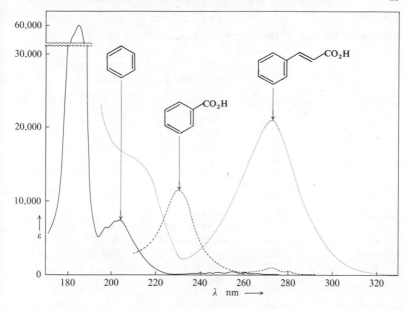

Fig. 1–5

originally at 254 nm, eventually overtaking it. This can be seen in the two other spectra recorded on Fig. 1–5: benzoic acid (the dashed line) shows the K-band at 230 nm with the B-band still clearly visible at 273 nm; but with the longer chromophore of cinnamic acid (dotted line) the K-band has moved to 273 nm and the B-band is completely submerged. In the latter case, we can see how the even stronger band, originally at 184 nm, has also moved, but has still not reached the accessible region. It is responsible for what is called end absorption; that is, the long-wavelength side of an absorption peak, the maximum of which is below the range of the instrument.

Table 1–9

Absorption Maxima of the Substituted Benzene Rings Ph—R

R	λ_{max} nm (ε) (solvent H_2O or MeOH)					
—H	203·5	(7400)	254	(204)		
—NH_3^+	203	(7500)	254	(160)		
—Me	206·5	(7000)	261	(225)		
—I	207	(7000)	257	(700)		
—Cl	209·5	(7400)	263·5	(190)		
—Br	210	(7900)	261	(192)		
—OH	210·5	(6200)	270	(1450)		
—OMe	217	(6400)	269	(1480)		
—SO_2NH_2	217·5	(9700)	264·5	(740)		
—CN	224	(13,000)	271	(1000)		
—CO_2^-	224	(8700)	268	(560)		
—CO_2H	230	(11,600)	273	(970)		
—NH_2	230	(8600)	280	(1430)		
—O^-	235	(9400)	287	(2600)		
—NHAc	238	(10,500)				
—COMe	245·5	(9800)				
—CH=CH_2	248	(14,000)	282	(750)	291	(500)
—CHO	249·5	(11,400)				
—Ph	251·5	(18,300)				
—OPh	255	(11,000)	272	(2000)	278	(1800)
—NO_2	268·5	(7800)				
—CH$\overset{t}{=}$CHCO$_2$H	273	(21,000)				
—CH$\overset{t}{=}$CHPh	295·5	(29,000)				

(Most values taken with permission from H. H. Jaffé and M. Orchin, *Theory and Applications of Ultraviolet Spectroscopy*, Wiley, New York, 1962.)

In disubstituted benzenes, two situations are important. (a) When electronically complementary groups, such as amino and nitro, are situated *para* to each other as in XII, there is a pronounced red shift in the main absorption band, compared to the effect of either substituent separately, due to the extension of the chromophore from the electron donating group to the electron withdrawing group through the benzene ring (XII, arrows). (b) Alternatively, when two groups are situated *ortho* or *meta* to each other or when the *para* disposed groups are not complementary, as in XIII, then the observed spectrum is usually closer to that of the separate, non-interacting, chromophores.

These principles are illustrated by the examples in Table 1–10.

The values in this table should be compared with each other and with the values for the single substituents separately given in Table 1–9.

λ_{max} 375 nm (ε 16,000) λ_{max} 260 nm (ε 13,000)

XII   XIII

In particular it should be noted that those compounds with non-complementary substituents, or with an *ortho* or *meta* substitution pattern, actually have a band (though a much weaker one) at longer wavelength than the compounds with interacting *para* disubstituted substituents. This fact is not in accord with the simple resonance picture; neither is the similarity of the *ortho* to the *meta* disubstituted cases. This is another case in which the molecular orbital theory (too complicated to be introduced here but dealt with in Murrell's book) gives a better picture.

Table 1–10

Absorption Maxima of the Disubstituted Benzene Rings R—C_6H_4—R′

R	R′	Orienta-tion	λ_{max}^{EtOH} in nm (ε)				
—OH	—OH	o-	214	(6000)	278	(2630)	
—OMe	—CHO	o-	253	(11,000)	319	(4000)	
—NH₂	—NO₂	o-	229	(16,000)	275	(5000)	405 (6000)
—OH	—OH	m-	277	(2200)			
—OMe	—CHO	m-	252	(8300)	314	(2800)	
—NH₂	—NO₂	m-	235	(16,000)	373	(1500)	
—Ph	—Ph	m-	251	(44,000)			
—OH	—OH	p-	225	(5100)	293	(2700)	
—OMe	—CHO	p-	277	(14,800)			
—NH₂	—NO₂	p-	229	(5000)	375	(16,000)	
—Ph	—Ph	p-	280	(25,000)			

In the case of disubstituted benzene rings in which the electron donating group is complemented by an electron withdrawing carbonyl group, some quantitative assessments may be made. These apply to the compounds R—C_6H_4—COX in which X is alkyl, H, OH, or OAlkyl, and refer to the strongest band in the accessible region; this is often the only measured band in the highly conjugated *para* disubstituted systems. The calculation is based on a parent value with increments for each substituent. Poly-substituted benzene rings should be treated with caution, particularly when the substitution might lead to steric hindrance preventing coplanarity of the carbonyl group and the ring. Table 1–11 gives the rules for this calculation. In the absence of steric hindrance to coplanarity, the calculated values are usually within 5 nm of the observed values.

Table 1–11

Rules for the Principal Band of Substituted Benzene Derivatives
$R—C_6H_4—COX$

	Orientation	λ_{calc}^{EtOH} nm
Parent Chromophore:		
X = alkyl or ring residue		246
X = H		250
X = OH or OAlkyl		230
Increment for each substituent:		
R = alkyl or ring residue	*o-, m-*	3
	p-	10
R = OH, OMe, OAlkyl	*o-, m-*	7
	p-	25
R = O^-	*o-*	11
	m-	20
	p-	78
R = Cl	*o-, m-*	0
	p-	10
R = Br	*o-, m-*	2
	p-	15
R = NH_2	*o-, m-*	13
	p-	58
R = NHAc	*o-, m-*	20
	p-	45
R = NHMe	*p-*	73
R = NMe_2	*o-, m-*	20
	p-	85

(Reprinted with permission from A. I. Scott, *Interpretation of the Ultraviolet Spectra of Natural Products*, Pergamon Press, Oxford, 1964.)

A single example, that of 6-methoxytetralone (XIV), will show the method.

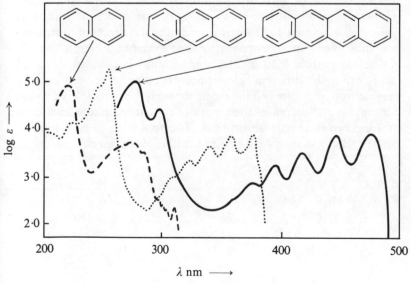

XIV

Parent value	246 nm
Ortho alkyl	3 nm
Para methoxyl	25 nm
λ_{calc}	274 nm

The maximum actually occurs at 276 nm (ε 16,500).

Other electron withdrawing groups, e.g., in nitriles and nitro compounds, show similar trends but with different and less well documented substituent effects.

1–21. Polycyclic Aromatic Hydrocarbons

The range of polycyclic aromatic hydrocarbons is too great for detailed consideration in this book. Their spectra are usually complicated, and for that reason are useful as fingerprints. This is

Fig. 1–6

(Reprinted with permission from R. A. Friedel and M. Orchin, *Ultraviolet Spectra of Aromatic Compounds*, Wiley, New York, 1951.)

particularly so in that the relatively non-polar substituents, such as alkyl and acetoxy groups, have only a small effect on the shape and position of the absorption peaks of the parent hydrocarbon. The degradation products of natural materials often contain polycyclic nuclei which can be identified in this way as, for example, a phenanthrene or a perylene. The spectra of a typical series, naphthalene, anthracene and naphthacene, are illustrated in Fig. 1–6; the logarithmic ordinate should be noted.

Fortunately, the collections of spectra mentioned in section 1–11 show the actual spectra of a great many of the known aromatic systems and make the identification of such systems a relatively simple matter.

1–22. Heteroaromatic Compounds

The range of heteroaromatic compounds is too great for detailed consideration in this book. In general they resemble the spectra of their corresponding hydrocarbons, but only in the crudest way. The heteroatom, whether like that in a pyrrole or that in a pyridine, leads to pronounced substituent effects which depend on the electron donating or withdrawing effect of the substituent and the heteroatom and on their orientation. The effects of these factors are predictable, in a qualitative way, using the same sorts of criteria as were used in section 1–20 when considering the effects of more than one substituent on a benzene ring. For example, a simple pyrrole XV and a pyrrole with an electron withdrawing substituent XVI have strikingly different absorption maxima. The conjugation present from the nitrogen lone-pair through the pyrrole ring to the carbonyl group increases the length of the chromophore and leads to longer wavelength absorption. The following illustrations of heterocyclic systems give some indication of the spectra observed.

XV

λ_{max}^{EtOH} 203 nm
(ε 5670)

XVI

λ_{max}^{EtOH} 262 nm
(ε 12,000)

XVII

λ_{max}^{EtOH} 245 nm
(ε 4800)

XVIII

λ_{max}^{EtOH} 300 nm
(ε 5000)

XIX

λ_{max}^{MeOH} 520 nm

XX

$\lambda_{max}^{CHCl_3}$ 245 nm (ε 12,000)
275 nm (ε 2800)
282 nm (ε 3020)

XXI

$\lambda_{max}^{cyclohexane}$ 220 nm (ε 26,000)
262 nm (ε 6310)
280 nm (ε 5620)
288 nm (ε 4170)

XXII

$\lambda_{max}^{CHCl_3}$ 218 nm (ε 79,000)
266 nm (ε 3900)
305 nm (ε 2000)
318 nm (ε 3000)
Compare these values
with the spectrum of
naphthalene in Fig. 1–6.

XXIII

λ_{max} pH 4   259·5 nm
pH 7   260 nm (ε 11,000)
pH 9·5 261 nm

XXIV

λ_{max} pH 1   210 nm (ε 9700)
276 nm (ε 10,000)
pH 5   269 nm (ε 6650)
pH 7   267 nm (ε 6130)
pH 12 272 nm (ε 5630)

	XXV	XXVI

λ_{max} pH 2   262 nm
        pH 7   260 nm (ε 13,500)
        pH 12 267 nm

λ_{max} pH 1   248 nm
                      271 nm
        pH 6   246 nm (ε 10,000)
               275 nm (ε 7800)
        pH 11 245 nm
               273 nm

In the case of potentially tautomeric molecules the change in the absorption maxima with the change of pH is due sometimes to a change in the chromophore as a result of the tautomerism and sometimes to simple protonation or deprotonation. This point is mentioned here in order to stress the importance of careful control of the medium in which spectra are taken. The changes in absorption maxima with change of pH are very useful diagnostically since they serve in some systems to identify the pattern of substitution. The stable tautomeric species have been identified, using ultraviolet spectroscopy. For example, the 2-hydroxypyridine (XXVII, R = H): pyrid-2-one (XXVIII, R = H) equilibrium has been shown to lie far to the right; the ultraviolet spectrum of the solution resembles that of a solution of N-methylpyrid-2-one (XXVIII, R = Me) and is different from that of 2-methoxypyridine (XXVII, R = Me).

XXVII	XXVIII
R = Me	R = Me

λ_{max} < 205 nm (ε > 5300)
        269 nm (ε 3230)

λ_{max} 226 nm (ε 6100)
        297 nm (ε 5700)

R = H
λ_{max} 224 nm (ε 7230)
        293 nm (ε 5900)

1–23. Quinones

A few representative quinones are illustrated below. The colour of the simpler members is due to the weak n→π* transition, similar to that of α-diketones.

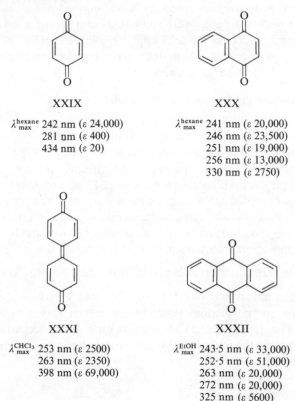

XXIX

$\lambda^{\text{hexane}}_{\text{max}}$ 242 nm (ε 24,000)
281 nm (ε 400)
434 nm (ε 20)

XXX

$\lambda^{\text{hexane}}_{\text{max}}$ 241 nm (ε 20,000)
246 nm (ε 23,500)
251 nm (ε 19,000)
256 nm (ε 13,000)
330 nm (ε 2750)

XXXI

$\lambda^{\text{CHCl}_3}_{\text{max}}$ 253 nm (ε 2500)
263 nm (ε 2350)
398 nm (ε 69,000)

XXXII

$\lambda^{\text{EtOH}}_{\text{max}}$ 243·5 nm (ε 33,000)
252·5 nm (ε 51,000)
263 nm (ε 20,000)
272 nm (ε 20,000)
325 nm (ε 5600)
405 nm (ε 90)

1–24. Porphyrins, Chlorins and Corroles

Our knowledge of the chemistry of these important groups of macrocyclic compounds has benefited considerably from the ease with which each class, in many of its various oxidation levels and with varying substitution patterns, can be recognized by the relative intensity of the four bands found in the visible region between 400 nm and 700 nm. In addition to these, a very strong sharp band (the Soret band) occurs near 400 nm (ε 100,000). It is interesting that another conjugated macrocyclic aromatic system, [18]-annulene, shows a similar intense band at 369 nm (ε 303,000).

These compounds are mentioned here to stress the importance and usefulness of ultraviolet and visible spectroscopy in the study of groups of compounds possessing a long, complicated chromophore. Although little can be accomplished in such systems from a theoretical point of view, the very large number of model systems available makes an empirical approach quite straightforward and very rewarding. These remarks apply to a large number of systems which, for one reason or another, have been studied, but which cannot be dealt with in this chapter.

1–25. Non-conjugated Interacting Chromophores

Non-conjugated systems usually have little effect on each other; diphenyl methane has a spectrum similar to that of toluene; the cross conjugation of the trieneone (X) was successfully ignored when calculating the expected absorption maximum; and even diphenyl ether is not very different from anisole. However, several special cases of non-conjugated interaction are known, two examples of which are given below. The unsaturated ketones (XXXIII) show the $n \rightarrow \pi^*$ and $\pi \rightarrow \pi^*$ transitions shifted in opposite directions when X becomes more electronegative. Presumably the π^* orbital is raised by transannular interaction with the $>\overset{+}{N}Me_2$ group, but since the n electron is closer to the $>\overset{+}{N}Me_2$ group in the excited state than in the ground state, the $n \rightarrow \pi^*$ transition is of lower energy. The diene (XXXIV) has absorption in the accessible ultraviolet whereas the isolated ethylenic double bond has no maximum above 190 nm.

XXXIII

X	λ_{max}	
$>CH_2$	238 nm	308·5 nm
$>\overset{+}{N}Me_2$	229 nm	318·5 nm

XXXIV

λ_{max} 205 nm (ε 2100)
214 nm (ε 1480)
220 nm (ε 870)
230 nm shoulder (ε 200)

1-26. The Effect of Steric Hindrance to Coplanarity

(*i*) *Steric hindrance to coplanarity about a double bond*, as in the hydrocarbon (**XXXV**), raises the ground state energy level but leaves the excited state relatively unchanged (the latter is probably of lowest energy in the conformation in which the biphenyl systems are at right angles). The result (in this case a series of bands culminating at 458 nm [ε 23,000]) is a shift toward the red from what might have been expected.

XXXV   XXXVI   XXXVII

(*ii*) *Mild steric hindrance to coplanarity about a single bond* has only a small effect on the position and intensity of absorption maxima.

(*iii*) *Medium steric hindrance to coplanarity about a single bond* gives rise to a marked decrease in intensity but may also lead to either a blue shift or a red shift. For example, the absorption maximum of the nitroaniline **XXXVI** (R = Me) is at 385 nm (ε 4840), showing a red shift and marked reduction in intensity from that of the parent compound **XXXVI** (R = H) at 375 nm (ε 16,000). Another example, in the opposite direction, is that of 2,4,6-trimethylacetophenone absorbing at 242 nm (ε 3200), which is to be compared with the calculated value (Table 1–11) of 262 nm and with *p*-methylacetophenone which has a maximum at 252 nm (ε 15,000).

(*iv*) *Extreme steric hindrance to coplanarity about a single bond* leads to a situation with no overlap between the separated chromophores. The dilactone (**XXXVII**) produced from shellolic acid showed no maximum in the accessible ultraviolet region but on hydrolysis of the $\alpha\beta$-unsaturated lactone grouping an acid with λ_{max} 227 nm (ε 5500) was obtained. This shows that the steric constraint of the lactone ring prevents conjugation and that release of this constraint then allowed the overlap of the double bond and carbonyl orbitals.

Bibliography

TEXTBOOKS

A. E. Gillam and E. S. Stern, *Electronic Absorption Spectroscopy*, Arnold, London, 2nd Ed., 1957.

C. N. R. Rao, *Ultraviolet and Visible Spectroscopy*, Butterworths, London, 1961.

A. I. Scott, *Interpretation of the Ultraviolet Spectra of Natural Products*, Pergamon Press, Oxford, 1964.

W. West (Ed.), *Technique of Organic Chemistry, Vol. IX, Chemical Applications of Spectroscopy*, Interscience, New York, 1956.

S. F. Mason, Chapter 7, The Electronic Absorption Spectra of Heterocyclic Compounds, *Physical Methods in Heterocyclic Chemistry, Vol. II*, Academic Press, New York, 1963.

THEORETICAL TREATMENTS

H. H. Jaffé and M. Orchin, *Theory and Applications of Ultraviolet Spectroscopy*, Wiley, New York, 1962.

J. N. Murrell, *The Theory of the Electronic Spectra of Organic Molecules*, Methuen, London, 1963.

G. R. Barrow, *Introduction to Molecular Spectroscopy*, McGraw-Hill, New York, 1962.

R. E. Dodd, *Chemical Spectroscopy*, Elsevier, Amsterdam, 1962.

E. F. H. Brittain, W. O. George, and C. H. J. Wells, *Introduction to Molecular Spectroscopy*, Academic Press, London, 1970.

See also p. 9 for catalogues of ultraviolet spectra.

2. Infrared Spectra

2–1. Introduction

The energy of most molecular vibrations corresponds to that of the infrared region of the electromagnetic spectrum. Molecular vibrations may be detected and measured either in an infrared spectrum or indirectly in a Raman spectrum. The most useful vibrations, from the point of view of the organic chemist, occur in the narrower range of 2·5 μ to 16 μ (1 μ = 10^{-4} cm.) which most infrared spectrometers cover. The position of an absorption band in the spectrum may be expressed in microns (μ), or very commonly —and throughout this book—in terms of the reciprocal of the wavelength, cm.$^{-1}$. The usual range of an infrared spectrum is, therefore, between 4000 cm.$^{-1}$ at the high frequency end and 625 cm.$^{-1}$ at the low frequency end.

Functional groups have vibration frequencies, characteristic of that functional group, within well defined regions of this range; these are summarized in Figs. 2–3, 2–4, 2–5 and 2–6, and form the subject matter of this chapter. We will, however, postpone discussion of the regions in which functional groups absorb until we have established the very great ease with which samples are prepared and spectra taken. The fact that many functional groups can be identified by their characteristic vibration frequencies makes the infrared spectrum the simplest, most rapid and often most reliable means for assigning a compound to its class.

2–2. Preparation of Samples and Examination in an Infrared Spectrometer

The spectrometer consists of a source of infrared light emitting radiation throughout the whole frequency range of the instrument. This light is split into two beams of equal intensity, and one beam is arranged to pass through the sample to be examined. If the frequency of a vibration of the sample molecule falls within the range of the instrument, the molecule may absorb energy of this frequency from the light. The spectrum is, therefore, scanned by comparing the intensity of the two beams after one has passed through the sample to be examined. The wavelength range over which the comparison is made is spread out in the usual way with a prism or grating. The whole operation is done automatically in such a way that the usual finished spectrum consists of a chart showing downward peaks, corresponding to absorption, plotted against wavelength or frequency. To allow for variations in the spectrometer, spectra are often calibrated against accurately known bands of the spectrum of polystyrene, the peaks of one or more of these bands being superimposed on the spectrum which is to be taken (see Fig. 2–2).

Compounds may be examined in the vapour phase, as pure liquids, in solution and in the solid state (see Fig. 2–13).

(*i*) *In the vapour phase.* The vapour is introduced into a special cell, usually about 10 cm. long, which can then be placed directly in the path of one of the infrared beams. The end walls of the cell are usually made of sodium chloride, which is transparent to infrared. Most organic compounds have too low a vapour pressure for this phase to be useful.

(*ii*) *As a liquid.* A drop of the liquid is squeezed between flat plates of sodium chloride (transparent throughout the 4000 to 625 cm.$^{-1}$ region). This is the simplest of all procedures.

(*iii*) *In solution.* The compound is dissolved, typically, to give a 1–5 per cent solution in carbon tetrachloride or, for its better solvent properties, chloroform freed from alcohol. This solution is introduced into a special cell, 0·1 to 1 mm. thick, made of sodium chloride. A second cell of equal thickness, but containing pure solvent, is placed in the path of the other beam of the spectrometer in order that solvent absorptions should be balanced. Spectra taken in such dilute solutions in non-polar solvents are generally the most desirable, because they are normally better resolved (see Fig. 2–13c) than spectra taken on solids, and also because intermolecular forces, which are especially strong in the crystalline state, are minimized. On the other hand, many compounds are not soluble in non-polar solvents, and all solvents absorb in the infrared; when the solvent absorption exceeds about 65 per cent of the incident light, spectra cannot be taken because insufficient light is transmitted to work the detection mechanism efficiently. Carbon tetrachloride and chloroform, fortunately, absorb over 65 per cent of the incident light only in those regions (Fig. 2–1) which are of little interest in diagnosis. Other solvents, of course, may be used but the areas of usefulness in each case should be checked beforehand, taking account of the size of the cell being used. In rare cases aqueous solvents are useful; special calcium fluoride cells are used.

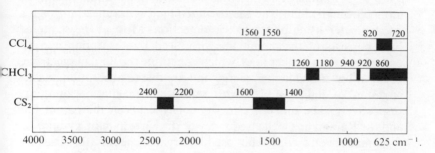

Fig. 2–1

Darkened areas are the regions in which the solvent cannot be used with a 0·2 mm. cell.

(*iv*) *In the solid state.* (a) About 1 mg. of a solid is finely ground in a small agate mortar with a drop of a liquid hydrocarbon (Nujol, Kaydol) or, if C—H vibrations are to be examined, with hexachlorobutadiene. The mull is then pressed between flat plates of sodium chloride. (b) Alternatively, the solid is ground with 10 to 100 times its bulk of pure potassium bromide and the mixture pressed into a disc using a special mould and a hydraulic press. The use of KBr eliminates the problem (usually not troublesome) of bands due to the mulling agent (see Figs. 2–13a and 2–13b) and tends, on the whole, to give rather better spectra, except that a band at 3450 cm.$^{-1}$, from the OH group of traces of water, almost always appears (see Fig. 2–9). Due to intermolecular interactions, band positions in solid state spectra are often different from those of the corresponding solution spectra. This is particularly true of those functional groups which take part in hydrogen bonding. On the other hand, the number of resolved lines is often greater in solid state spectra (see Figs. 2–13), so that comparison of the spectra of, for example, synthetic and natural samples in order to determine identity, is best done in the solid state. This is only true, of course, when the same crystalline modification is in use; racemic synthetic material and optically active natural material, for example, should be compared in solution.

2–3. Examination in a Raman Spectrometer

Raman spectra are generally taken on machines using laser sources, and the quantity of material needed is now of the order of a few mg. A liquid or a concentrated solution is irradiated with the monochromatic light, and the *scattered* light is examined through a spectrometer using photoelectric detection. Most of the scattered light consists of the parent line produced by absorption and re-emission. Much weaker lines, which constitute the Raman spectrum, occur at lower and higher energy and are due to absorption and re-emission of light coupled with vibrational excitation or decay respectively. The difference in frequency between the parent line and the Raman line is the frequency of the corresponding vibration.

Raman spectroscopy is not used by organic chemists for structure determination as routinely as is infrared spectroscopy, but for the detection of certain functional groups (see section 2–4), and for the *analysis* of mixtures—of deuterated compounds for example—it has found much use.

2–4. Selection Rules

Infrared light is absorbed when the oscillating dipole moment (due to a molecular vibration) interacts with the oscillating electric vector of the infrared beam. A simple rule for deciding if this interaction (and hence absorption of light) occurs is that the dipole moment at one extreme of a vibration must be different from the dipole moment at the other extreme of the vibration. In the Raman effect a corresponding interaction occurs between the light and the molecule's polarizability, resulting in different selection rules.

The most important consequence of these selection rules is that in a molecule with a centre of symmetry those vibrations sym-metrical about the centre of symmetry are active in the Raman and inactive in the infrared; those vibrations which are not centro-symmetric are inactive in the Raman and usually active in the infrared. This is doubly useful, for it means that the two types of spectrum are complementary; and the more easily obtained, the infrared, is the most informative for organic chemists, because most functional groups are not centrosymmetric.

The symmetry properties of a molecule in a solid can be different from those of an isolated molecule. This can lead to the appearance of infrared absorption bands in a solid state spectrum which would be forbidden in solution or in the vapour phase.

2–5. The Infrared Spectrum

A complex molecule has a large number of vibrational modes which involve the whole molecule. To a good approximation, however, some of these molecular vibrations are associated with the vibrations of individual bonds or functional groups (localized vibrations) while others must be considered as vibrations of the whole molecule.

The localized vibrations are either stretching, bending, rocking, twisting or wagging. For example, the localized vibrations of the methylene group are

| Symmetric stretching | Asymmetric stretching | Bending or scissoring | Rocking | Twisting | Wagging |

Many localized vibrations are very useful for the identification of functional groups.

The soggy vibrations of the molecule as a whole give rise to a series of absorption bands at low energy, below 1500 cm.$^{-1}$, the positions of which are characteristic of that molecule. These bands make those localized vibrations which have frequencies below 1500 cm.$^{-1}$ less useful for diagnostic purposes since confusion of one with the other may occur. Frequently bands are observed which do not correspond to any of the fundamental vibrations of the molecule and are due to overtone bands and combination bands, the latter as a result of interaction between two or more vibrations. Occasionally these bands are useful diagnostically but more usually they supplement the region below 1500 cm.$^{-1}$ The net result, when a spectrum has been taken, is a region above 1500 cm.$^{-1}$ showing absorption bands assignable to a number of functional groups and a region, characteristic of the compound in question and no other compound, containing many bands below 1500 cm.$^{-1}$ This region, for obvious reasons, is called the fingerprint region. The use of the fingerprint region to confirm the identity of a compound with an authentic sample is considerably more reliable, in most cases, than the technique of taking a mixed melting point. Within the fingerprint region some bands assignable to functional groups do occur and may be used diagnostically; such identifications should be regarded as helpful rather than as definitive.

The regions in which functional groups absorb are summarized in Fig. 2–2, which also illustrates the very simple spectrum of the liquid paraffin Nujol, the mulling agent often used when taking the

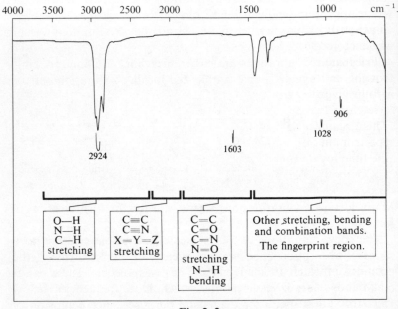

Fig. 2–2

spectrum of a solid sample. Since Nujol possesses only C—H and C—C bonds, its spectrum shows features found in the majority of organic compounds.

The stretching vibrations of single bonds to hydrogen give rise to the absorption at the high frequency end of the spectrum as a result of the low mass of the hydrogen atom. Thereafter, the order of stretching frequencies follows the order: triple bonds at higher frequency than double bonds and double bonds higher than single bonds—on the whole the greater the strength of the bond between two similar atoms the higher the frequency of the vibration. Bending vibrations are of much lower frequency and usually appear in the fingerprint region below 1500 cm.$^{-1}$ An exception is the N—H bending vibration which appears in the 1600 cm.$^{-1}$ region.

2–6. The Use of the Tables of Characteristic Group Frequencies

Each of the three high frequency ranges above 1500 cm.$^{-1}$ shown in Fig. 2–2 is expanded in the four charts Figs. 2–3, 2–4, 2–5 and 2–6. These charts summarize the narrower ranges within which each of the functional groups absorbs. The absorption bands which are found in the fingerprint region and which are assignable to functional groups are summarized in the chart Fig. 2–7; these latter correlations are occasionally useful, either because they are sometimes strong bands in otherwise featureless regions or because their absence may rule out incorrect structures. Following these summaries are Tables 2–1 to 2–21 arranged by functional groups roughly in order of their stretching frequencies. Where a functional group gives rise to absorption bands in addition to those due to stretching frequencies, their position is also mentioned in the table. This organization enables one to examine the main regions of the spectrum in turn, rather than to work backwards from a guess as to the functional groups present.

One could deal with the spectrum of an unknown as follows. Examine each of the three main regions of the spectrum covered by Figs. 2–3 to 2–6; at this stage certain combinations of structures can be ruled out and some tentative conclusions reached. Where there is still ambiguity, the tables corresponding to those groups which might be present should be consulted, whereupon more detailed information should be available. It is well to be sure that the bands under consideration are of the appropriate intensity for the structure suspected. Some assistance in this task is provided at the end of this chapter, where nine infrared spectra are reproduced in order to show the usual appearance of a number of characteristically shaped bands.

Except where otherwise stated, band positions are given for dilute solution in non-polar solvents. Intensities in the infrared are less frequently recorded and less conveniently measured than is the case with ultraviolet spectra. Usually intensities are expressed subjectively as strong (s), medium (m), weak (w) and, in books such as this, variable (v). The position of all bands is given in cm.$^{-1}$

2–7. Correlation Charts

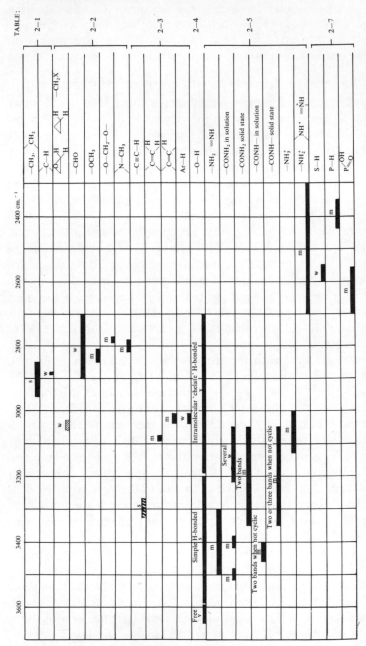

Fig. 2–3. Stretching frequencies of single bonds to hydrogen. (Hatched areas are those with less well defined limits.)

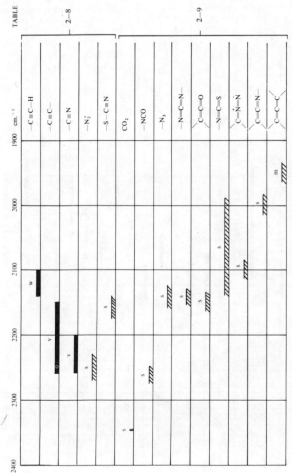

Fig. 2–4. Stretching frequencies of triple bonds and cumulated double bonds. (Hatched areas are those with less well defined limits.)

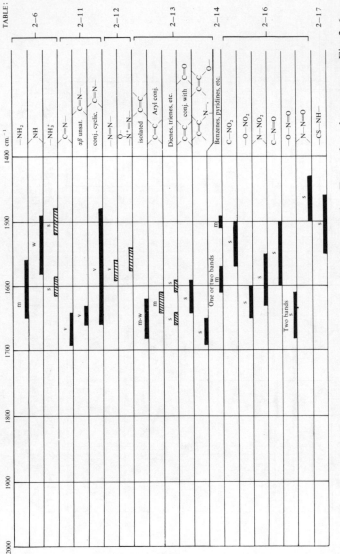

Fig. 2–5. Double bond stretching and N—H bending frequencies. For carbonyl groups see Fig. 2-6. (Hatched areas are those with less well defined limits.)

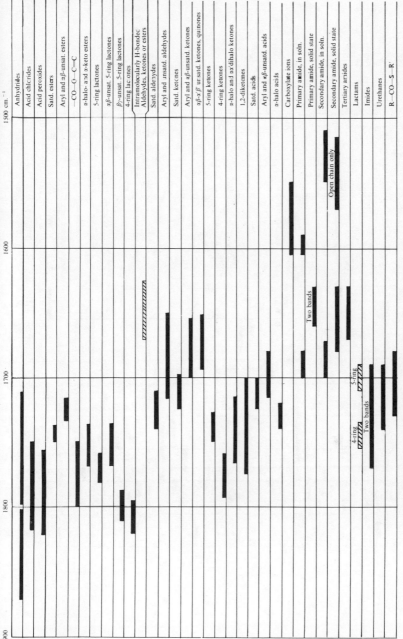

Fig. 2–6. Stretching frequencies of carbonyl groups. All values are found in Table 2–10. All bands are strong.

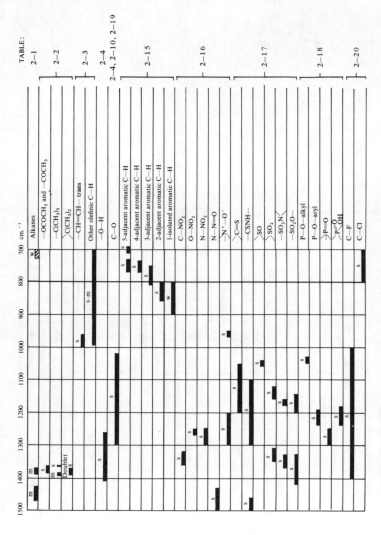

Fig. 2-7. Some characteristic bands in the fingerprint region.

2–8. Absorption Frequencies of Single Bonds to Hydrogen
Table 2–1
Saturated C—H and C—C

Group	Band	Remarks
$>\text{CH}_2$ —CH$_3$	2960–2850(s)	Two or three bands usually; $>$C—H stretching
$>$CH	2890–2880(w)	
$>\text{CH}_2$ —CH$_3$	1470–1430(m)	$>$C—H deformations
—CH$_3$	1390–1370(m)	—CH$_3$ symmetrical deformation
$>\text{CH}_2$	~720(w)	$>$CH$_2$ rocking

Table 2–2
Miscellaneous C—H

Group	Band	Remarks
Cyclopropane C—H Epoxide C—H —CH$_2$-halogen	~ 3050(w)	C—H stretching; cf. alkenes
—CO—CH$_3$	3100–2900(w)	Often very weak
—CHO	2900–2700(w)	Usually two bands, one near 2720 cm.$^{-1}$ (see Fig. 2–13a)
—O—CH$_3$	2850–2810(m)	
—O—CH$_2$—O—	2790–2770(m)	
N—CH$_3$ and N—CH$_2$—	2820–2780(m)	(see Fig. 2–12)
—C(CH$_3$)$_3$	1395–1385(m) 1365(s)	
$>$C(CH$_3$)$_2$	~ 1380(m)	A roughly symmetrical doublet (see Fig. 2–8)
—O—CO—CH$_3$ —CO—CH$_3$	1385–1365(s) 1360–1355(s)	The high intensity of these bands often dominates this region of the spectrum

Table 2–3

Alkene and Aromatic C—H

See also Table 2–13 and Table 2–14 for the corresponding double bond absorptions, and Table 2–15 for the aromatic C—H out-of-plane bending vibrations.

Group	Band	Remarks
—C≡C—H	~3300(s)	
$>$C=C$<$ H, H	3095–3075(m)	C—H stretching; sometimes obscured by the much stronger bands of saturated C—H groups which occur below 3000 cm.$^{-1}$
$>$C=C$<$ H	3040–3010(m)	(See Fig. 2–12)
Aryl—H	3040–3010(w)	Often obscured (but see Fig. 2–9)
R, H $>$C=C$<$ H, R	970–960(s)	C—H out-of-plane deformation. When the double bond is conjugated with, for example, a C=O group this band is shifted towards 990 cm.$^{-1}$
RCH=CH$_2$	995–985(s) and 940–900(s)	
R$_2$C=CH$_2$	895–885(s)	
R$_2$C=C$<$ H, R	840–790(m)	
H, R $>$C=C$<$ H, R	730–675(m)	

Much is known about the precise position of the various CH, CH$_2$ and CH$_3$, symmetrical and unsymmetrical vibration frequencies. C—H bonds do not take part in hydrogen bonding and, therefore, their position is little affected by the state of measurement or their chemical environment. C—C vibrations, which absorb in the fingerprint region, are generally weak and not practically useful. Since most organic molecules possess alkane residues, the groups of saturated C—H absorption bands given above are of little diagnostic value; their general appearance may be seen in the Nujol spectrum (Fig. 2–2). The absence of saturated C—H absorption in a spectrum is, of course, diagnostic evidence for the absence of such a part structure in the corresponding compound (see Fig. 2–13a). Unsaturated and aromatic C—H stretching frequencies can be distinguished from the saturated C—H absorption since the latter occurs below 3000 cm.$^{-1}$ (see, however, Table 2–2) while the former gives rise to much less intense absorption above 3000 cm.$^{-1}$ Alkene and aromatic C—H absorption is covered by Table 2–3.

A few special structural features in saturated C—H groupings give rise to characteristic absorption bands. These are summarized in Table 2–2 on p. 47.

Table 2–4

Alcohol and Phenol —O—H

Group	Band	Remarks
Water in solution	3710	
Free —OH	3650–3590(v)	Sharp; O—H stretching (see Fig. 2–8)
H-bonded —OH (solid, liquid and dilute solution)	3600–3200(s)	Often broad but may be sharp for some intramolecular single bridge H-bonds; the lower the frequency the stronger the H-bond (see Fig. 2–8)
Intramolecular H-bonded —OH in chelate form (see also Table 2–10, carboxylic acids)	3200–2500(v)	Broad; the lower the frequency the stronger the H-bond; sometimes so broad as to be overlooked
Water of crystallization (solid state spectra)	3600–3100(w)	Usually a weak band at 1640–1615 cm.$^{-1}$ also; water in trace amounts in KBr discs shows a broad band at 3450 cm.$^{-1}$ (see Fig. 2–9)
—O—H	1410–1260(s)	O—H bending
>C—OH	1150–1040(s)	C—O stretching (see Fig. 2–8)

The value of the O—H stretching frequency has been used for many years as a test for and measure of the strength of hydrogen bonds. The stronger the hydrogen bond the longer the O—H bond, the lower the vibration frequency and the broader and the more intense the absorption band. The sharp free 'monomeric' band in the 3650–3590 cm.$^{-1}$ range can be observed in the vapour phase, in dilute solution or when such factors as steric hindrance prevent hydrogen bonding. Pure liquids, solids and many solutions show only the broad, 'polymeric' band in the 3600–3200 range. Frequently liquid phase spectra show both bands (see Fig. 2–8).

Intramolecular hydrogen bonds of the non-chelate type, for example in 1,2-diols, show a sharp band in the range 3570–3450 cm.$^{-1}$, the precise position being a measure of the strength of the hydrogen bond. A similar, though rather less sharp band is observed when the hydrogen bonding gives rise to dimers only. The 'polymeric' band is generally much broader. Distinctions can be made among the various possibilities by testing the effect of dilution; intramolecular hydrogen bonds are unaffected and the absorption band is, therefore, unaffected; while intermolecular hydrogen bonds are broken, leading to a decrease in the bonded O—H absorption and an increase in or the appearance of free O—H absorption. Spectra taken of samples in the solid state show only the broad strong band in the range 3400–3200 cm.$^{-1}$.

The stretching frequencies of N—H bonds (see Table 2–5) can sometimes be confused with those of hydrogen bonded O—H frequencies. Due to their much weaker tendency to form hydrogen bonds, N—H absorption is usually sharper; moreover, N—H absorption is of weaker intensity and, in dilute solutions, it never gives rise to absorption as high as the free O—H range near 3600 cm.$^{-1}$ Weak bands, overtones of the strong carbonyl absorption in the 1800–1600 cm.$^{-1}$ region, also appear in the 3600–3200 cm.$^{-1}$ region.

The effects of hydrogen bonding can be seen when a carbonyl group is the acceptor, for its characteristic stretching frequency is also lowered (see Table 2–10).

The characteristic series of bands in the 3000–2500 cm.$^{-1}$ range produced by most carboxylic acids can be seen in Fig. 2–9. The highest frequency band is due to the O—H stretching vibration, and the other bands to combination vibrations. The bands are usually seen as a jagged series on the low frequency side of any C—H absorption which may be present. Combined with a carbonyl absorption in the correct region (Table 2–10) this series is very useful for the identification of carboxylic acids.

Table 2–5
Amine, Imine, Ammonium, and Amide N—H
N—H Stretching

Group	Band	Remarks
Amine and imine $>$N—H $=$N—H	3500 3300(m)	Primary amines show two bands in this range; the unsymmetrical and symmetrical stretching. Secondary amines absorb weakly. The pyrrole and indole N—H band is sharp (see Fig. 2–10)
—NH_3^+ Amino acids	3130–3030(m)	Values for solid state; broad; bands also (but not always) near 2500 and 2000 cm.$^{-1}$ (see text below Fig. 2–10)
Amino salts	$\sim$3000(m)	
$>$$NH_2^+$ $>$NH$^+$ $=\overset{+}{N}H$	2700–2250(m)	Values for solid state; broad, due to the presence of overtone bands, etc.
Primary amide —$CONH_2$	$\sim$3500(m) $\sim$3400(m)	Lowered $\sim$150 cm.$^{-1}$ in the solid state and on H-bonding; often several bands 3200–3050 cm.$^{-1}$ (see Fig. 2–11)
Secondary amide —CONH—	3460–3400(m)	Two bands; lowered on H-bonding and in the solid state (see Fig. 2–13a). Only one band with lactams
	3100–3070(w)	A weak extra band with bonded and solid state samples (see Fig. 2–13a)

Much is known of amide N—H absorptions; the appearance of two bands being ascribed to forms I and II. The carbonyl region of many amides (Table 2–10) also shows two bands.

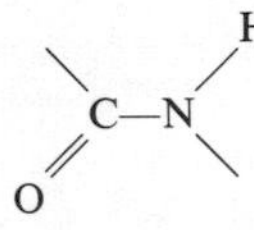

I   II

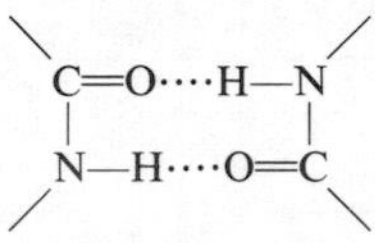

Hydrogen bonding lowers and broadens N—H stretching frequencies to a lesser extent than was the case with O—H groups. The intensity of N—H absorption is usually less than that of O—H absorption.

Table 2–6

N—H **Bending**

See also Table 2–10 for amide absorptions in this region.

Group	Band	Remarks
—NH$_2$	1650–1560(m)	
$\rangle$NH	1580–1490(w)	Often too weak to be noticed
—NH$_3^+$	1600(s) 1500(s)	Secondary amine salts have the 1600 cm.$^{-1}$ band

Table 2–7

Miscellaneous R—H

Group	Band	Remarks
—S—H	2600–2550(w)	Weaker than O—H and less affected by H-bonding
P—H	2440–2350(m)	Sharp
P$\langle^O_{OH}$	2700–2560(m)	Associated OH
R—D	1/1·37 times the corresponding R—H frequency	Useful when assigning R—H bands, deuteration leading to a known shift to lower frequency

2–9. Absorption Frequencies of Triple Bonds and Cumulated Double Bonds

Table 2–8

Triple Bonds

Group	Band	Remarks
—C≡C—H	3300(m)	C—H stretching
	2140–2100(w)	C≡C stretching
—C≡C—	2260–2150(v)	* † ‡ (see Fig. 2–9)
—C≡N	2260–2200(v)	C≡N stretching; stronger and to the lower end of the range when conjugated; occasionally very weak (see Fig. 2–14) or absent; for example some cyanohydrins show no C≡N absorption
Diazonium salts R—$\overset{+}{N}$≡N	~2260	
Thiocyanates R—S—C≡N	2175–2140(s)	Aryl thiocyanates at upper end of the range, alkyl at the lower end

* Conjugation with olefinic or acetylenic groups lowers the frequency and raises the intensity. Conjugation with carbonyl groups usually has little effect on the position of absorption.

† Symmetrical and nearly symmetrical substitution makes the C≡C stretching frequency inactive in the infrared. It is, however, seen clearly in the Raman spectrum.

‡ When more than one acetylenic linkage is present, and sometimes when there is only one, there are frequently more absorption bands in this region than there are triple bonds to account for them.

Table 2–9
Cumulated Double Bonds

Group	Band	Remarks
Carbon dioxide O═C═O	2349(s)	Appears in many spectra due to inequalities in path length
Isocyanates —N═C═O	2275–2250(s)	Very high intensity; position unaffected by conjugation
Azides —N$_3$	2160–2120(s)	
Carbodiimides —N═C═N—	2155–2130(s)	Very high intensity; split into an unsymmetrical doublet by conjugation with aryl groups
Ketenes $>$C═C═O	~2150(s)	
Isothiocyanates —N═C═S	2140–1990(s)	Broad and very intense
Diazoalkanes R$_2$C═$\overset{+}{N}$═$\overset{-}{N}$	~2100(s)	
Ketenimines C═C═N—	~2000(s)	
Allenes C═C═C	~1950(m)	Two bands when terminal allene or when bonded to electron attracting groups, e.g., —CO$_2$H

The ranges quoted in Table 2–9 are tentative, since relatively few compounds in some of these classes have been examined.

The unusually high double bond frequencies encountered in the X═Y═Z systems are believed to arise from strong coupling of the two separate stretching vibrations, the asymmetrical and symmetrical stretching frequencies becoming widely separated. This type of coupling occurs only when two groups with similar high frequency vibrations and the same symmetry are situated near one another. Other examples in which such coupling is found are the amide group (Table 2–10) and the carboxylate ion (Table 2–10).

2–10. The Aromatic Overtone and Combination Region, 2000–1600 cm.$^{-1}$

Aromatic compounds are characterized by the weak C—H stretching band near 3030 cm.$^{-1}$ (Table 2–3) and by bands near 1600 and 1500 cm.$^{-1}$ (Table 2–14). Occasionally, the substitution pattern on a benzene ring can be deduced from the strong bands associated with the C—H out of plane bending vibrations below 900 cm.$^{-1}$ (see Table 2–15). In addition to these bands, there are bands in the 1225–950 cm.$^{-1}$ region which are of little use, and a group of weak overtone and combination bands in the 2000–1600 cm.$^{-1}$ region. It has been found from spectra taken with more concentrated solutions that the shape and number of the two to six bands found in this region is a function of the substitution pattern of the benzene ring. The use of this region, however, depends on the absence of other absorption in the region. The characteristic patterns for the various substituted benzenes are given in Nakanishi's book.

2–11. Absorption Frequencies of the Double Bond Region

Table 2–10

Carbonyl Absorption $>C\!=\!O$ *All bands quoted are strong.*

Groups	Band	Remarks
Acid anhydrides —CO—O—CO—		
Saturated	1850–1800 1790–1740	Two bands usually separated by about 60 cm.$^{-1}$. The higher frequency band is more intense in acyclic anhydrides and the lower frequency band is more intense in cyclic anhydrides
Aryl and $\alpha\beta$-unsaturated	1830–1780 1770–1710	
Saturated five-ring	1870–1820 1800–1750	
All classes	1300–1050	One or two strong bands due to C—O stretching
Acid chlorides —COCl		
Saturated	1815–1790	Acid fluorides higher, bromides and iodides lower
Aryl and $\alpha\beta$-unsaturated	1790–1750	
Acid peroxides —CO—O—O—CO—		
Saturated	1820–1810 1800–1780	
Aryl and $\alpha\beta$-unsaturated	1805–1780 1785–1755	

Table 2–10 *continued*

Groups	Band	Remarks
Esters and lactones —CO—O—		
Saturated	1750–1735	
Aryl and $\alpha\beta$-unsaturated	1730–1715	
Aryl and vinyl esters C=C—O—CO—Alkyl	1800–1750	The C=C stretching band also shifts to higher frequency
Esters with electronegative α-substituents; e.g., $>$CCl—CO—O—	1770–1745	
α-keto esters	1755–1740	
Six-ring and larger lactones	Similar values to the corresponding open chain esters	
Five-ring lactone	1780–1760	
$\alpha\beta$-unsaturated five-ring lactone	1770–1740	When α-C—H present there are two bands, the relative intensity depending on the solvent (see Fig. 2–14)
$\beta\gamma$-unsaturated five-ring lactone; i.e., vinyl ester type	~1800	
Four-ring lactone	~1820	
β-keto ester in H-bonding enol form	~1650	Keto form normal; chelate type H-bond causes shift to lower frequency than the normal ester. The C=C is usually near 1630(s) cm.$^{-1}$
All classes	1300–1050	Usually two strong bands due to C—O stretching.
Aldehydes —CHO (see also Table 2–2 for C—H). All values given below are lowered in liquid film or solid state spectra by about 10–20 cm.$^{-1}$. Vapour phase spectra have values raised about 20 cm.$^{-1}$		
Saturated	1740–1720	
Aryl	1715–1695	(See Fig. 2–13a) *Ortho* hydroxy or amino groups shift this value to 1655–1625 cm.$^{-1}$ due to intramolecular H-bonding
$\alpha\beta$-unsaturated	1705–1680	
$\alpha\beta,\gamma\delta$-unsaturated	1680–1660	
β-ketoaldehyde in enol form	1670–1645	Lowering caused by chelate type H-bonding

Table 2–10 *continued*

Groups	Band	Remarks
Ketones C=O All values given below are lowered in liquid film or solid state spectra by about 10–20 cm.$^{-1}$. Vapour phase spectra have values raised about 20 cm.$^{-1}$		
Saturated	1725–1705	
Aryl	1700–1680	
$\alpha\beta$-unsaturated	1685–1665	
$\alpha\beta,\alpha'\beta'$-unsaturated and diaryl	1670–1660	
Cyclopropyl	1705–1685	
Six-ring ketones and larger	Similar values to the corresponding open chain ketones	(See Fig. 2–8)
Five-ring ketones	1750–1740	$\alpha\beta$-unsaturation, etc., has a similar effect on these values as on those of open chain ketones
Four-ring ketones	~1780	
α-halo ketones	1745–1725	Affected by conformation; highest values are obtained when both halogens are in the same plane as the C=O
α,α'-dihalo ketones	1765–1745	
1,2-Diketones s-*trans*: (i.e., open chains)	1730–1710	Antisymmetrical stretching frequency of both C=O's. The symmetrical stretching is inactive in the infrared but active in the Raman
1,2-Diketones s-*cis*, six-ring	1760 and 1730	
1,2-Diketones s-*cis*, five-ring	1775 and 1760	
o-Amino- or *o*-hydroxy-aryl ketones	1655–1635	Low due to intramolecular H-bonding. Other substituents and steric hindrance etc. affect the position of the band
Quinones	1690–1660	C=C usually near 1600(s) cm.$^{-1}$
Extended quinones	1655–1635	
Tropone	1650	Near 1600 cm.$^{-1}$ when lowered by H-bonding as in tropolones
Carboxylic acids —CO$_2$H		
All types	3000–2500	O—H stretching; a characteristic group of small bands due to combination bands etc. (For the appearance of this group see Fig. 2–9)

Table 2–10 *continued*

Groups	Band	Remarks
Saturated	1725–1700	The monomer is near 1760 cm.$^{-1}$ but is rarely observed. Occasionally both bands, the free monomer and the H-bonded dimer, can be seen in solution spectra. Ether solvents give one band near 1730 cm.$^{-1}$
$\alpha\beta$-unsaturated	1715–1690	(See Fig. 2–9)
Aryl	1700–1680	
α-halo-	1740–1720	
Carboxylate ions —CO_2^- For amino acids, see text below Fig. 2–10		
Most types	1610–1550 1420–1300	Antisymmetrical and symmetrical stretching respectively
Amides —CO—N$\langle$ (See also Table 2–5 and 2–6 for N—H stretching and bending) Primary —$CONH_2$		
In solution	~1690	Amide I; C=O stretching
Solid state	~1650	
In solution	~1600	Amide II; mostly N—H bending
Solid state	~1640	
		Amide I is generally more intense than amide II. (In the solid state amide I and II may overlap.) (See Fig. 2–11)
Secondary —CONH—		
In solution	1700–1670	Amide I (see Fig. 2–13)
Solid state	1680–1630	
In solution	1550–1510	Amide II; found in open chain amides only (see Fig. 2–13)
Solid state	1570–1515	
		Amide I is generally more intense than amide II
Tertiary	1670–1630	Since H-bonding is absent solid and solution spectra are much the same (see Fig. 2–12)
Lactams		
Six- and larger rings	~1670	
Five-ring	~1700	Shifted to higher frequency when the N atom is in a bridged system
Four-ring	~1745	
R—CO—N—C=C		Shifted +15 cm.$^{-1}$ by the additional double bond

Table 2-10 *continued*

Groups	Band	Remarks
C=C—CO—N		Shifted by up to $+15$ cm.$^{-1}$ by the additional double bond. This is an unusual effect for $\alpha\beta$-unsaturation. It is said to be due to the inductive effect of the C=C on the well conjugated CO—N system, the usual conjugation effect being less important in such a system
Imides —CO—N—CO—		
Cyclic six-ring	~ 1710 and ~ 1700	Shift of $+15$ cm.$^{-1}$ with $\alpha\beta$-
Cyclic five-ring	~ 1770 and ~ 1700	unsaturation
Ureas N—CO—N		
RNHCONHR	~ 1660	
Six-ring	~ 1640	
Five-ring	~ 1720	
Urethanes		
R—O—CO—N	1740–1690	Also shows amide II band when non- or mono-substituted on N
Thioesters and Acids RCO—S—R′		
RCOSH	~ 1720	$\alpha\beta$-unsaturated or aryl acid or ester shifted ~ -25 cm.$^{-1}$
RCOS—alkyl	~ 1690	
RCOS—aryl	~ 1710	

Intensities of carbonyl bands. Acids generally absorb more strongly than esters, and esters more strongly than ketones or aldehydes. Amide absorption is usually similar in intensity to that of ketones but is subject to much greater variations.

Position of carbonyl absorption. The general trends of structural variation on the position of C=O stretching frequencies may be summarized as follows:

(*a*) The more electronegative the group X in the system R—CO—X—, the higher is the frequency.

(*b*) $\alpha\beta$-unsaturation causes a lowering of frequency of 15–40 cm.$^{-1}$, except in amides, where little shift is observed and that usually to higher frequency.

(*c*) Further conjugation has relatively little effect.

(d) Ring strain in cyclic compounds causes a relatively large shift to higher frequency. This phenomenon provides a remarkably reliable test of ring size, distinguishing clearly between four, five and larger membered ring ketones, lactones and lactams. Six-ring and larger ketones, etc., show the normal frequency found for the open chain compounds.

(e) Hydrogen bonding to a carbonyl group causes a shift to lower frequency of 40–60 cm.$^{-1}$. Acids, amides, enolized β-keto carbonyl systems and o-hydroxy- and o-aminophenyl carbonyl compounds show this effect. All carbonyl compounds tend to give slightly lower values for the carbonyl stretching frequency in the solid state compared with the value for dilute solutions.

(f) Where more than one of the structural influences on a particular carbonyl group is operating, the net effect is usually close to additive.

Table 2–11

Imines, Oximes etc. $>C{=}N{-}$

Group	Band	Remarks
$>C{=}N{-}H$	3400–3300(m)	N—H stretching; lowered on H-bonding
$>C{=}N{-}$	1690–1640(v)	Difficult to identify due to large variations in intensity and the closeness to C=C stretching region. Oximes usually give very weak bands
$\alpha\beta$-unsaturated	1660–1630(v)	
Conjugated cyclic systems	1660–1480(v)	

Table 2–12

Azo Compounds $-N{=}N{-}$

Group	Band	Remarks
$-N{=}N{-}$	~ 1575(v)	Very weak or inactive in infrared. Sometimes seen in Raman
$-\overset{+}{N}{=}N{-}$ $\,\,\mid$ O^-	~ 1570	

Table 2–13

Alkenes $>$C$=$C$<$

(See also Table 2–3 for the $=$C—H absorptions of alkenes.)

Group	Band	Remarks
Non-conjugated $>$C$=$C$<$	1680–1620(v)	May be very weak if more or less symmetrically substituted (see Fig. 2–12)
Conjugated with aromatic ring	~1625(m)	More intense than with unconjugated double bonds
Dienes, trienes, etc.	1650(s) and 1600(s)	Lower frequency band usually more intense and may hide or overlap the higher frequency band
$\alpha\beta$-unsaturated carbonyl compounds	1640–1590(s)	Usually much weaker than the C$=$O band (see, however, Fig. 2–11)
Enol esters, enol ethers and enamines	1690–1650(s)	(See Fig. 2–14)

The most substituted double bonds tend to absorb at the high frequency end of the range, the least substituted at the low frequency end. The absorption may be very weak when the double bond is more or less symmetrically substituted, but the vibration frequency can then be detected and measured in the Raman spectrum. For the same reason *trans*-double bonds tend to absorb less strongly than *cis*-double bonds. Table 2–3 should be consulted for the $=$C—H vibration frequencies, which may give additional structural information.

A general trend which has been observed is the effect caused by strain on the stretching frequency of double bonds. A double bond exocyclic to a ring shows the same pattern as cyclic ketones: that is the frequency rises as the ring size decreases. A double bond within a ring shows the opposite trend: that is, the frequency falls as the ring size decreases. The C—H stretching frequency rises slightly as ring strain increases.

Table 2–14

Aromatic Compounds

See also Table 2–3 and Table 2–15 for aryl—H vibration frequencies.

Group	Band	Remarks
Aromatic rings	~1600(m)	(See Fig. 2–13)
	~1580(m)	Stronger when the ring is further conjugated
	~1500(m)	This is usually the strongest of the two or three bands

The two or three bands in the 1600–1500 cm.$^{-1}$ region are shown by most six-membered aromatic ring systems such as benzenes, polycyclic systems and pyridines. They constitute a valuable identification of such a system. Further bands are shown by aromatic rings in the fingerprint region between 1225 and 950 cm.$^{-1}$ which are of little diagnostic value. The weak overtone and combination bands in the 2000 to 1660 cm.$^{-1}$ region have been mentioned on p. 55. A fourth group of bands below 900 cm.$^{-1}$ is produced by the out-of-plane C—H bending vibrations (see Table 2–15).

Table 2–15

Substitution Patterns of the Benzene Ring

Group	Band	Remarks
Five adjacent H	770–730(s) and 720–680(s)	Mono-substituted
Four adjacent H	770–735(s)	*Ortho*-disubstituted (see Fig. 2–10)
Three adjacent H	810–750(s)	*Meta*-disubstituted etc. and 1, 2, 3-trisubstituted
Two adjacent H	860–800(s)	*Para*-disubstituted etc. (see Fig. 2–13a)
Isolated H	900–800(w)	*Meta*-disubstituted etc.; usually not strong enough to be useful

The frequency of the C—H out-of-plane vibration is determined by the number of adjacent hydrogen atoms on the ring and hence the frequency is a means of determining the substitution pattern. This does not work as well in practice as one might hope. These strong bands are not always the only—or even the strongest—bands in the region (for example, C-halogen frequencies interfere particularly) so that assignments based on this evidence alone should be treated with caution. For example, the spectrum of tryptophan (Fig. 2–10) and the spectrum of p-acetamidobenzaldehyde (Fig. 2–13a) show only the characteristic absorption of *ortho-* and *para-* disubstituted benzene rings respectively; but the spectrum of p-nitrophenylpropiolic acid (Fig. 2–9) shows not only a band at 860 cm.$^{-1}$, consistent with its being a p-disubstituted benzene, but also bands at 750 and 685 cm.$^{-1}$, consistent with its being monosubstituted. It is obvious that a positive assignment of substitution pattern is not possible in the latter case.

The values in Table 2–15 hold reasonably well for condensed ring systems and for pyridines. Powerful electron withdrawing substituents tend to shift the values to higher frequency.

Table 2–16

Nitro, Nitroso, etc. N=O

Group	Band	Remarks
C—NO_2	~ 1560(s) ~ 1350(s)	Lowered ~ 30 cm.$^{-1}$ when conjugated. The two bands are due to asymmetrical and symmetrical stretching of the NO bonds (see Fig. 2–9)
Nitrates O—NO_2	1650–1600(s) 1270–1250(s)	
Nitramines N—NO_2	1630–1550(s) 1300–1250(s)	
C—N=O	1600–1500(s)	
O—N=O	1680–1610(s)	Two bands
N—N=O	1500–1430(s)	
$>\overset{+}{N}$—$\overset{-}{O}$ aromatic aliphatic	1300–1200(s) 970–950(s)	Very strong bands
NO_3^-	1410–1340 860–800	

2.12. Groups Absorbing in the Fingerprint Region

Table 2–17

Sulphur Compounds

Group	Band	Remarks
—S—H	2600–2550(w)	S—H stretching; weaker than O—H and less affected by H-bonding. This absorption is strong in the Raman
$>C=S$	1200–1050(s)	
$\overset{\displaystyle >}{\underset{\displaystyle \parallel S}{C}}-N<$	~3400	N—H stretching; lowered to ~3150 cm.$^{-1}$ in the solid state
	1550–1460(s)	Amide II
	1300–1100(s)	Amide I
$>S=O$	1060–1040(s)	
$>SO_2$	1350–1310(s)	
	1160–1120(s)	
$-SO_2-N<$	1370–1330(s)	
	1180–1160(s)	
$-SO_2-O-$	1420–1330(s)	
	1200–1145(s)	

Table 2–18

Phosphorus Compounds

Group	Band	Remarks
P—H	2440–2350(s)	Sharp
P—Ph	1440(s)	Sharp
P—O-alkyl	1050–1030(s)	
P—O-aryl	1240–1190(s)	
P=O	1300–1250(s)	
P—O—P	970–910	Broad
$P<^O_{OH}$	2700–2560	H-bonded O—H
	1240–1180(s)	P=O stretching

Table 2–19

Ethers

Group	Band	Remarks
$>$C—O—C$<$	1150–1070(s)	C—O stretching
$=$C—O—C$<$	1275–1200(s) 1075–1020(s)	
C—O—CH$_3$	2850–2810(m)	C—H stretching; aryl ethers at higher end of the range
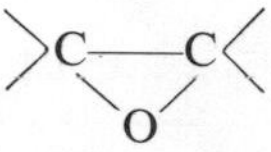 $>$C———C$<$ $\backslash$O$/$	~1250 ~900 ~800	

Table 2–20

Halogen Compounds

Group	Band	Remarks
C—F	1400–1000(s)	
C—Cl	800–600(s)	
C—Br	750–500(s)	
C—I	~500(s)	

Table 2–21

Inorganic Ions

Group	Band	Remarks
Ammonium	3300–3030	All bands strong
Cyanide, Thiocyanate, Cyanate	2200–2000	
Carbonate	1450–1410	
Sulphate	1130–1080	
Nitrate	1380–1350	
Nitrite	1250–1230	
Phosphates	1100–1000	

2–13. Examples of Infrared Spectra

The following spectra show the appearance and relative intensities of the absorption peaks due to a number of functional groups. The wide variety of fingerprints shows the usefulness of this region for identification.

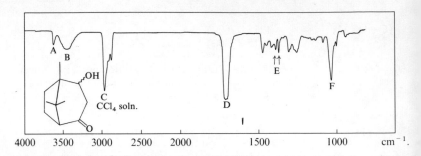

Fig. 2–8

A	3620 cm.$^{-1}$	Free O—H
B	3460 cm.$^{-1}$	Intermolecular and weakly bonded O—H
C	2960 cm.$^{-1}$	Saturated C—H
D	1710 cm.$^{-1}$	Ketone C=O
E	1370 and 1390 cm.$^{-1}$	$\underset{CH_3}{\overset{CH_3}{>C<}}$
F	1035 cm.$^{-1}$	C—O stretching

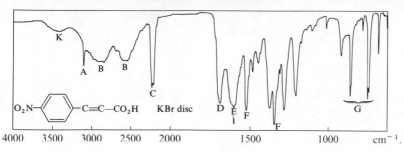

Fig. 2–9

A	3100 cm.$^{-1}$	Aryl C—H stretching
B	3200–2400 cm.$^{-1}$	Characteristic strongly H-bonded O—H of carboxylic acid
C	2225 cm.$^{-1}$	Conjugated C≡C, hence strong
D	1690 cm.$^{-1}$	Conjugated —CO_2H
E	1605 cm.$^{-1}$	Benzene ring, unusually broad and un-resolved. A band near 1500 cm.$^{-1}$ is masked
F	1520 and 1350 cm.$^{-1}$	Conjugated nitro group—NO_2
G	950–650 cm.$^{-1}$	An example of a substituted ring in which it is not possible to decide with any certainty, due to the large number of bands in the region, in favour of 1, 4-disubstitution. For an example where the assignment can be made with confidence see Fig. 2–10
K		OH of water almost always present even in good KBr discs like this one

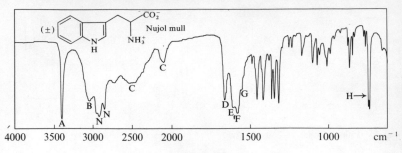

Fig. 2–10

A   3400 cm.$^{-1}$ Indole N—H
B   3040 cm.$^{-1}$ Broad 'ammonium' band due to —NH_3^+
C   ~2500 and   Two bands, very common with amino acids;
    ~2100 cm.$^{-1}$ also shown by primary amine salts
D   1665 cm.$^{-1}$ Amino acid I; unusually strong
E   1610 cm.$^{-1}$ Possibly aryl group
F   1590 cm.$^{-1}$ Amino acid II; the ionized carboxylate group
                            —CO_2^-
G   1550 cm.$^{-1}$ —NH_3^+ deformations
H   750 or 740 cm.$^{-1}$ C—H out-of-plane deformations showing a
                            1, 2-disubstituted benzene ring
N   Nujol peaks

Amino acids show the spectrum of the zwitterionic groups. The primary ammonium —NH_3^+ group N—H stretching appears under the peaks of the saturated C—H absorption. The two bands near 2500 and 2000 cm.$^{-1}$ are frequently found when the —NH_3^+ group is present, and are due to overtones and combinations. In the double bond region, there are several peaks, including at least one due to N—H bending and one, the strongest, due to the ionized carboxyl group. The highest frequency N—H bending peak is often very weak.

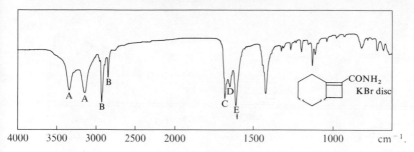

Fig. 2–11

A	3340 and 3140 cm.$^{-1}$	Typical amide —NH$_2$ pair of bands
B	2840 and 2930 cm.$^{-1}$	Saturated C—H
C	1680 cm.$^{-1}$	Amide I
D	1650 cm.$^{-1}$	Amide II
E	1610 cm.$^{-1}$	Conjugated and strained C=C

This spectrum shows the pair of N—H stretching bands and the pair of bands in the C=O region typical of primary amides in the solid state. The C=C peak appears here as an unusually strong peak. The amide I and II bands are not always so well resolved in the solid state.

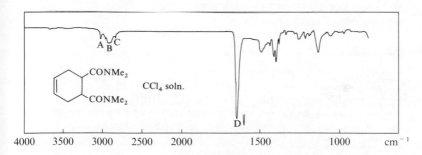

Fig. 2–12

A	3020 cm.$^{-1}$	Olefinic C—H stretch
B	2920 cm.$^{-1}$	Saturated C—H stretch
C	2830 cm.$^{-1}$	N—CH$_3$ C—H stretch
D	1650 cm.$^{-1}$	Tertiary amide C=O

This spectrum shows the absence of N—H, the strong sharp C=O of a tertiary amide, and, because of the symmetry of the molecule, no C=C stretching absorption.

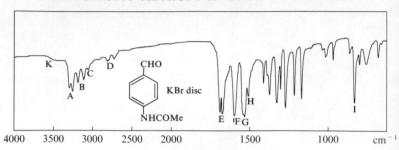

Fig. 2–13a

A	3300 and 3260 cm.$^{-1}$	Secondary amide N—H
B	3190 and 3110 cm.$^{-1}$	Secondary amide bands of unknown origin
C	3060 cm.$^{-1}$	Aryl C—H
D	2810 and 2730 cm.$^{-1}$	Aldehyde C—H
E	1695 and 1680 cm.$^{-1}$	Aldehyde C=O and amide I
F	1600 cm.$^{-1}$	Benzene ring
G	1535 cm.$^{-1}$	Amide II
H	1510 cm.$^{-1}$	Benzene ring
I	835 cm.$^{-1}$	p-Disubstituted benzene ring
K		Shoulder of OH band from traces of water in the KBr

This spectrum shows the multiplicity of bands found with secondary amides. The presence of so many bands in the spectra of such compounds as secondary amides is probably caused by the many ways in which such groupings can associate with each other, of which those shown on p. 51 are only two of many possibilities.

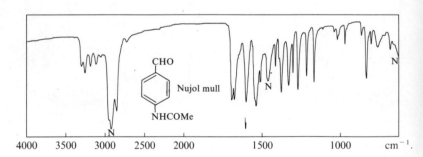

Fig. 2–13b

This spectrum was taken on the same compound as that in Fig. 2–13a, but was done as a Nujol mull. The main features are closely similar, but it can be seen how one of the Nujol peaks (labelled N) can obscure an important peak such as one of those labelled D on Fig. 2–13a. On the other hand, the band marked K on Fig. 2–13a is no longer present.

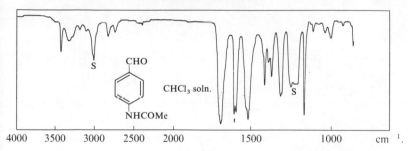

Fig. 2–13c

This is again a spectrum of the same compound, but taken in solution. This time some changes in appearance are apparent. The N—H stretch region is markedly different, and the amide I band has moved to a slightly higher frequency, making it coincident with the aldehyde C=O band. These changes are expected of a secondary amide, where a change in the nature of the intermolecular associations occurs in going from the solid state into the solution state. Such changes most affect the vibration frequencies of the functional groups involved in those associations.

The benzene ring band near 1600 cm.$^{-1}$ is now resolved into the two bands often found here: solution-state spectra are often better resolved than solid-state spectra. On the other hand there are many more strong lines in the fingerprint region in the solid state spectra.

The bands marked S are partly due to the solvent, because the absorption of the solvent has been incompletely cancelled (see p. 37).

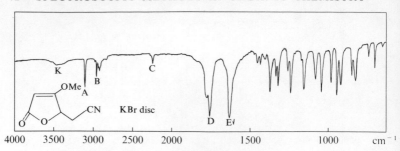

Fig. 2–14

A	3100 cm.$^{-1}$	Vinyl C—H stretch
B	2960–2900 cm.$^{-1}$	Saturated C—H stretch
C	2250 cm.$^{-1}$	Unconjugated C≡N stretch
D	1770 and 1755 cm.$^{-1}$	α,β-Unsaturated-γ-lactone C=O stretch
E	1630 cm.$^{-1}$	Vinyl ether C=C stretch

This spectrum shows how weak the unconjugated C≡N absorption can be, and how $\alpha\beta$-unsaturation, which lowers the frequency of the vibrations of the carbonyl group, combines with the presence of a five-membered ring, which raises the frequency, to give a band at 1755 cm.$^{-1}$, near the normal position of saturated esters and six-membered ring lactones. The extra band at 1770 cm.$^{-1}$ is common with $\alpha\beta$-unsaturated five-membered ring lactones having an α—H.

Bibliography

TEXTBOOKS

L. J. Bellamy, *The Infrared Spectra of Complex Molecules*, Methuen, London, 3rd Ed. 1966.

L. J. Bellamy, *Advances in Infrared Group Frequencies*, Methuen, London, 1969.

A. D. Cross, *Introduction to Practical Infrared Spectroscopy*, Butterworths, London, 2nd Ed. 1964.

K. Nakanishi, *Infrared Absorption Spectroscopy*, Holden-Day, San Francisco, 1962.

J. H. van der Maas, *Basic Infrared Spectroscopy*, Heyden, 2nd Ed., London 1972.

C. N. R. Rao, *Chemical Applications of Infrared Spectroscopy*, Academic Press, New York, 1963.

A. R. Katritzky and A. P. Ambler, Chapter 10, Infrared Spectra, *Physical Methods in Heterocyclic Chemistry*, Vol. II, Academic Press, New York, 1963.

K. Nakamoto, *Infrared Spectra of Inorganic and Coordination Compounds*, Wiley, New York, 1963.

H. A. Szymanski, *Interpreted Infrared Spectra*, Plenum, New York, 3 Vols., 1964.

F. S. Parker, *Applications of Infrared Spectroscopy in Biochemistry, Biology and Medicine*, Hilger, London, 1971.

J. Loader, *Basic Laser Raman Spectroscopy*, Heyden, London, 1970.

T. R. Gilson and P. J. Hendra, *Laser Raman Spectroscopy*, Wiley, London, 1970.

H. A. Szymanski, Ed., *Raman Spectroscopy: Theory and Practice*, Plenum, New York, Vol. 1, 1967, Vol. 2, 1970.

THEORETICAL TREATMENTS

G. Herzberg, *Infrared and Raman Spectra of Polyatomic Molecules*, Van Nostrand, Princeton, 1945.

G. R. Barrow, *Introduction to Molecular Spectroscopy*, McGraw-Hill, New York, 1962.

R. P. Bauman, *Absorption Spectroscopy*, Wiley, New York, 1962.

R. E. Dodd, *Chemical Spectroscopy*, Elsevier, Amsterdam, 1962.

G. W. King, *Spectroscopy and Molecular Structure*, Holt, Rinehart and Winston, New York, 1964.

R. Zbinden, *Infrared Spectroscopy of High Polymers*, Academic Press, New York, 1964.

Mansel Davies (Ed.) *Infrared Spectroscopy and Molecular Structure*, Elsevier, Amsterdam, 1963.

R. G. J. Miller and B. C. Stace, *Laboratory Methods in Infrared Spectroscopy*, Heyden, London, 2nd Ed., 1972.

CATALOGUES OF SPECTRA

H. M. Hershenson, *Infrared Absorption Spectra*, Academic Press, New York, Index for 1945–1957 (1959); Index for 1958–1962 (1964).

An Index of Published Infrared Spectra, H.M.S.O., London, Vols. I and II, 1960.

Sadtler Standard Spectra (Infrared Prism and Infrared Grating), Heyden, London, 1970. A collection of 39,000 and 19,000 spectra respectively; additions are made yearly.

The Aldrich Library of Infrared Spectra, Aldrich Chemical Co., Milwaukee, 1971. An ordered collection of 8000 spectra.

Documentation of Molecular Spectroscopy, Butterworths, London. A collection of spectra on coded cards, elaborately cross indexed.

3. Nuclear Magnetic Resonance Spectra

3–1. Note

Many of the NMR spectra illustrated in this chapter are reproduced from the Varian NMR catalogues, volumes 1 and 2, with the permission of Varian Associates to whom we express our most sincere thanks. Each spectrum is referenced in terms of its catalogue number. These spectra have been determined in dilute solution (*ca.* 7 per cent) in $CDCl_3$ and at 60 MHz.

3–2. Nuclear Spin and the Spectrometer

The NMR phenomenon (first observed in 1946) is observable because certain nuclei behave like tiny spinning bar magnets. Most important among such nuclei are ^1H, ^{13}C, ^{19}F and ^{31}P, all having nuclear spin values of $\frac{1}{2}$. Certain other nuclei which are important in organic chemistry have a nuclear spin value of zero and therefore give no nuclear resonance signals; these include ^{12}C and ^{16}O.

If a proton is placed in a uniform magnetic field, it may take up one of two orientations with respect to the field; these may be considered to be: a low-energy orientation in which the nuclear magnet is aligned with the field, and a high-energy orientation in which it is aligned against the field. The transition between these two energy states can be brought about by the absorption of a quantum of suitable electromagnetic radiation of energy hν. It turns out that if we employ field strengths of the order of 10,000 gauss, the energy required to flip the nuclear magnet is supplied by the radiofrequency range of the electromagnetic spectrum (in practice around 10 to 100 MHz).

The simple outline given in the previous paragraph indicates that to observe nuclear magnetic resonance signals for protons, we will require:

    (*i*) a radio-frequency transmitter
    (*ii*) a homogeneous magnetic field
    (*iii*) a radio-frequency receiver

It will be seen in section 3–3 that since different protons in an organic molecule have varying electronic environments, the precise value of the magnetic field required to bring any one into resonance at constant frequency will vary slightly from proton to proton. Thus if we operate the NMR spectrometer at a fixed frequency, a fourth requirement in the instrument will be a unit to sweep the magnetic field over a small range. A schematic diagram showing the relationship between these basic components and the sample is given in Fig. 3–1.

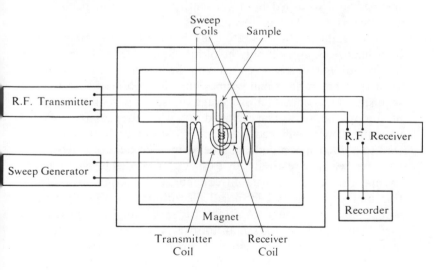

Fig. 3–1

In practice, the field sweep required to bring protons occurring in the vast majority of chemical environments into resonance is only ten parts in a million. This fact places stringent requirements on the homogeneity of the magnetic field employed in NMR experiments. These requirements may be met in part by designing the magnet with a much smaller air gap between the poles than has in fact been shown in the diagram (Fig. 3–1). Moreover, by spinning the solution of the sample about the vertical axis of the sample tube, field inhomogeneities perpendicular to the direction of spinning can be averaged out.

In practice, the most difficult part in obtaining a good NMR spectrum lies in obtaining a homogeneous field in the region of the sample. Once this has been achieved, the remaining operations are relatively simple. The sample (as little as 1 mg. in the case of some modern instruments, but preferably about 10–30 mg.) is dissolved in the solvent (most frequently about 0·5–1·0 ml. of carbon tetrachloride or deuterochloroform ($CDCl_3$)) and the resulting solution transferred to the sample tube, which, having an internal diameter of about 4 mm., will be filled to a depth of approximately 3–7 cm. A few per cent of a reference substance (tetramethylsilane, see section 3–3) is then added to the solution, the sample tube is placed between the poles of the magnet, and the spectrum may be scanned and recorded by a conventional chart and pen-and-ink recorder within a few minutes.

If the available sample size is a few hundred microgrammes, then the signal response will only be of the same order of magnitude as the 'noise' inherently associated with the instrument. The spectrum must then be scanned many times and the information from each scan stored in a computer of average transients (CAT). Noise, being of random nature, will be averaged out, while the signals associated with the sample will add up. In this way, repeated scanning can lead to a good spectrum from a very small amount of material.

So far, we have referred only to spectra recorded by continuously sweeping the field at constant frequency. A serious disadvantage of this approach [resulting in so-called continuous wave (cw) spectra] is that only a very small portion of the spectrum is being excited at any given time. Thus, the signal/noise (S/N) ratio which is attainable in a given time is adversely affected, and this situation is particularly undesirable if we are dealing with a nucleus where sensitivity is a problem. Such is the case for ^{13}C, where the low natural abundance (1·1 per cent) and magnetogyric ratio render the sensitivity about 10^{-4} of that observed for ^1H. For such nuclei, or indeed when sample size is a problem (e.g., 0·1–1 mg. for ^1H spectra), Fourier transform techniques can advantageously be employed. The radio frequency is applied at one end of the spectrum as a short, powerful pulse and this behaves like a spread of frequencies. If Δ Hz is the entire range of chemical shifts to be recorded in the spectrum, then the pulse length (t_p sec.) must be chosen such that

$$t_p \ll \tfrac{1}{4}\Delta$$

Pulse lengths for ^{13}C spectra are of the order of μsecs.

All nuclei in the spectrum can be excited by a single pulse and, as they decay back to their equilibrium state, the receiver coils of the spectrometer record a decay of magnetization which takes the form of a complex series of sine waves decaying exponentially in time. This information is related to the normal NMR spectrum *via* a Fourier transform which can conveniently be carried out by an on-line computer. Many thousands of pulsed spectra (each *spectrum* being recorded in seconds) can be accumulated before the Fourier transform operation is carried out, and the result is a dramatic improvement in S/N for a given expenditure of time.

Pulsed spectra and Fourier transform techniques are most commonly used to obtain ^{13}C spectra (section 3–13). Unless otherwise stated the spectra reproduced in this chapter are obtained by continuous wave techniques.

3–3. Chemical Shift

The frequency (v) at which any proton will resonate in the NMR spectrum is given by equation 3–1, in which H is the local field experienced by the proton and γ is a constant known as the magneto-gyric ratio.

$$v = \frac{\gamma H}{2\pi} \tag{3–1}$$

The local field (H) experienced by the proton will not correspond to the applied magnetic field (H_0) since the nucleus will in general be shielded by the electrons surrounding it. The extent of this shielding may be represented in terms of a shielding parameter (σ), defined such that the local field (H) is given by equation 3–2.

$$H = H_0(1-\sigma), \tag{3–2}$$

therefore

$$v = \frac{\gamma H_0(1-\sigma)}{2\pi} \tag{3–3}$$

Equation 3–3 follows and hence it may be seen that protons with different shielding parameters, i.e., different electronic environments, may be successively brought into resonance either by a frequency sweep of the spectrum at a constant field strength, or by a field sweep of the spectrum at a constant frequency. A complete proton magnetic resonance spectrum will have been obtained for a given compound when equation 3–3 has been satisfied for every proton in the molecule. Of course, it is not possible to tell by inspection of a given NMR spectrum whether it was obtained by a sweep of the frequency or of the magnetic field.

The positions of proton resonances in an NMR spectrum are measured relative to the resonance position of the twelve equivalent protons of an arbitrary reference substance, tetramethylsilane (TMS, I, see Fig. 3–2). The twelve protons of this molecule, being

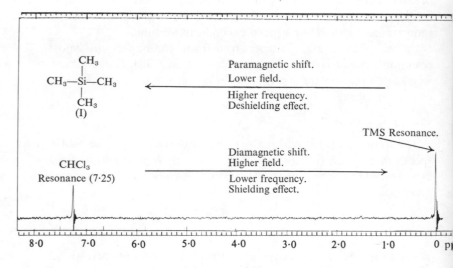

Fig. 3–2

chemically equivalent, all resonate at the same value of the applied field and therefore give rise to a single line (Fig. 3–2). TMS makes a convenient reference substance for the following reasons.

(*i*) It is a volatile liquid which may be added in trace amounts to a solution of the sample in carbon tetrachloride or deuterochloroform ($CDCl_3$). Recovery of the pure sample, if necessary, is usually then a simple matter.

(*ii*) Protons in the vast majority of organic compounds resonate at lower field than the protons of TMS. Therefore, by arbitrarily assigning $v_{TMS} = 0$, we can define a scale such that most proton resonances will be of the same sign (positive, for convenience).

(*iii*) TMS does not readily become involved in intermolecular associations with the sample. Such intermolecular associations would be undesirable, since they would modify the electronic environment of the TMS protons and hence change the absolute resonance position of these protons.

The NMR spectrum could, in principle, be calibrated either in cycles per second (Hz—frequency units) or in milligauss (field units), though conventionally 'Hz' is employed. Although we are measuring $v_S - v_{TMS}$, where v_S and v_{TMS} are resonance frequencies of the sample and tetramethylsilane, respectively, it is not always convenient to express our results in Hz, because it is evident from equation 3–3 that the frequency of any given resonance will be proportional to the applied magnetic field. Since several models of spectrometers are available which employ different field strengths, a scale of field-independent units must be chosen. A parameter δ has been defined by equation 3–4 and obviously δ will be field-independent because the operating frequency of an instrument is directly proportional to the strength of the magnetic field.

$$\delta = \frac{v_S - v_{TMS}}{\text{Operating freq. in MHz}} \qquad (3\text{–}4)$$

The numerator in this equation is expressed in Hz, as opposed to the denominator which is expressed in MHz and therefore *the units of δ, the chemical shift parameter, are parts per million (ppm)*. In the instance of a 60 MHz spectrometer, equations 3–5 and 3–6 follow.

$$\delta = \frac{v_S - v_{TMS}}{60} \text{ ppm}, \qquad (3\text{–}5)$$

therefore, since we arbitrarily assign $v_{TMS} = 0$,

$$\delta = \frac{v_S}{60} \text{ ppm} \qquad (3\text{–}6)$$

A number of terms are in common use to describe the position of the resonance of one proton relative to another, or the movement of a peak in the spectrum, and these terms are summarized in Fig. 3–2. It is important to appreciate the following two points.

(*i*) The frequency of a *particular resonance* increases in direct proportion to the increase in the field strength. This information permits us to compare spectra obtained on instruments operating at different field strengths (e.g., 60 MHz and 100 MHz instruments).

(*ii*) In any *one spectrum* which must be determined at *either* constant frequency *or* constant field, frequency increases in the direction of decreasing magnetic field strength, as indicated in Fig. 3–2.

The figure (Fig. 3–2) is, in fact, the NMR spectrum of a dilute solution of chloroform and TMS in deuterochloroform. Small amounts of chloroform almost invariably contaminate the deutero-chloroform employed in NMR work and so it is important to associate the peak at $\delta = 7.25$ ppm with this solvent. Deutero-chloroform is very popular for NMR studies because it is an excellent solvent and, like carbon tetrachloride, contains no protons which might obscure regions of the spectrum.

An alternative parameter which is frequently employed to describe the chemical shift of a proton is τ, which is related to δ as indicated in equation 3–7. Both δ and τ are currently in common use, but we shall use only δ in this chapter.

$$\tau = 10 - \delta \text{ ppm} \qquad (3-7)$$

A. Intramolecular Factors affecting Chemical Shift

It is now reasonable to inquire about the factors which determine the location of a proton resonance in the spectrum. In dilute solution the factors which influence the chemical shift parameter (δ) are very predominantly intramolecular; these may be divided into three main categories.

(*i*) *Inductive effect.* If an atom is placed in a uniform magnetic field, the electrons surrounding the nucleus are caused to circulate in a manner producing a secondary field which is opposed to the applied field in the region of the nucleus (II). The circulation of electrons therefore causes shielding of the nucleus.

H_0

II, ● = nucleus
• = electron

CH_3—$\overset{|}{\underset{|}{C}}$—   ($\delta \sim 0.9$ ppm)

CH_3—$\overset{|}{N}$—   ($\delta \sim 2.3$ ppm)

CH_3—O—   ($\delta \sim 3.3$ ppm)

III

However, if the electron density round an atom is reduced due to the inductive effect of a neighbouring electronegative atom, the secondary field is decreased, with the result that resonance can now occur at a lower value of the applied field. Hence electron withdrawal from a proton by an electronegative atom causes deshielding of the proton. For example, the protons of a methyl group attached to oxygen resonate at lower field than those of a methyl group attached to nitrogen, which in turn resonate at lower field than the protons of a C-methyl group (see III). It is emphasized that the quoted δ values are approximate only, since the precise value of the chemical shift may be affected by more distant structural variations.

(ii) *Anisotropy of chemical bonds.* Generally speaking, the modification of the applied magnetic field in the region of a given structural unit of a molecule (e.g., a carbonyl group, IV) will be both distance (r) and angle (θ) dependent. This arises because the applied field can cause circulations of the electrons within a functional group, but the circulations may be more facile in one plane than another. The electron circulations are therefore angle dependent, and hence so are the induced secondary magnetic fields. Thus chemical bonds are magnetically anisotropic and it will be seen below that the chemical shift of a proton is frequently modified by neighbouring functional groups. The anisotropy of the carbonyl group causes deshielding ($-$) of protons lying in a cone extending from the carbonyl oxygen atom, but shielding ($+$) of protons lying outside this cone (see V).

IV   V   VI

It should be noted that anisotropy effects are field effects operating through space, in contrast to inductive effects which operate through the chemical bonds. An additional field effect may arise due to the dipole moment of a functional group. To illustrate the operation of field effects we may refer to the NMR spectrum of pulegone (VI), in which methyl group 'a' resonates at $\delta = 1\cdot77$ ppm and 'b' at $\delta = 1\cdot95$ ppm. Obviously in the absence of the carbonyl group, these resonances would be coincident; the difference in chemical shift induced by the carbonyl group could have been attributed either to an inductive or to a field effect. The latter explanation is the correct one, 'b' being deshielded because of the anisotropy and dipolar effects of the keto-group; the inductive effect would be practically negligible through four bonds.

(*iii*) *Van der Waals deshielding*. It is an experimental fact that if protons attached to different atoms are brought sufficiently close together for van der Waals repulsion to operate, mutual deshielding of the protons will occur. Hence the proton H* in a conformationally rigid cyclohexanone chair system (VII, embedded in a steroid skeleton for example) will resonate at lower field when R = CH_3 than when R = H. The paramagnetic shifts caused by intramolecular van der Waals repulsion are usually of the order of 1 ppm or less. Steric interaction causes *deshielding* only, since the effective shielding of the hydrogen nucleus is decreased on asymmetrical distortion of the electron cloud (associated with mutual repulsion of the electron clouds—see VIII, which is a view of VII from in front of the carbonyl group).

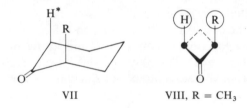

VII   VIII, R = CH_3

In summary, the inductive effect consistently leads to paramagnetic shifts, which may be several ppm in magnitude. Likewise, steric congestion will result in deshielding, but the influence will normally be less than 1 ppm. The field effects due to the anisotropy and dipole moment of a chemical moiety are important in determining both long and short range shifts, the signs and magnitudes of which are dependent on the distances and angles involved.

In Table 3–1, which is appended to this chapter, the chemical shifts of methyl, methylene and methine protons in frequently encountered environments are given. In the vast majority of cases, the shifts are largely determined by the inductive effects of the electronegative atoms. Obviously, steric and anisotropy factors, which are usually specific properties for each individual molecule considered, cannot be included in such a table. However, the peaks usually appear within $\pm0\cdot2$ ppm of the values quoted, unless anisotropic and steric effects of nearby groups are operating. It is evident that for any given functional group the situation δ(methyl) < δ(methylene) < δ(methine) usually prevails.

The chemical shifts of protons adjacent to two $(X_1—CH_2—X_2)$

$$X_2$$
$$|$$

or three $(X_1—CH—X_3)$ functional groups can be estimated by the application of *Shoolery's rules*. An effective shielding constant (C_i) for X_i is derived empirically by averaging the chemical shifts caused by successive substitution of methane by X_i. The constants so obtained (Table 3–2, end of chapter) are then added to the chemical shift of methane ($0\cdot23$ ppm):

$$\delta_{CH_2} = 0\cdot23 + C_1 + C_2 \qquad \delta_{CH} = 0\cdot23 + C_1 + C_2 + C_3$$

The shielding constants are a useful guide to the relative shielding powers of functional groups. The additivity principle has its limitations (especially for calculating δ_{CH}), but gives results within $0\cdot6$ ppm of experimental values in most cases (e.g. $\delta_{calc.}^{CHCl_3} = 7\cdot73$ ppm; $\delta_{obs.}^{CHCl_3} = 7\cdot25$ ppm). For the cases of trisubstitution, errors in excess of 1 ppm occur with a frequency of about 20–30 per cent; if the observed and calculated values differ by more than $2\cdot5$ ppm, the proposed substitution pattern can probably be ruled out.

B. Effect of Concentration, Solvent and Temperature
(Nitrogen Quadrupole Relaxation and Deuterium Exchange)

In the concentration range $0\cdot05$ to $0\cdot5$ molar in deuterochloroform or carbon tetrachloride solution, the resonance positions of most protons attached to carbon are barely influenced (±1 Hz at 60 MHz). Similarly, the influence of temperature on such protons is not usually important. In contrast to the behaviour of protons attached to carbon, the line positions of those bonded directly to electronegative atoms (e.g., in OH, SH and NH groups) exhibit a marked dependence on concentration and temperature, for reasons which are outlined in the following paragraph.

If the spectrum of a compound is determined in both carbon tetrachloride and deuterochloroform solutions, a comparison of the traces reveals only very small variations in chemical shift ($\pm0\cdot1$ ppm, that is ±6 Hz at 60 MHz) for C—H protons in most cases. Again the protons of OH, SH and NH groups may show much larger solvent dependencies. The reason for the sensitivity of the chemical shift of protons in OH, SH and NH groups towards changes in concentration, temperature and solvent may be partially understood from a consideration of the NMR spectra of acetic acid and water. In the spectrum of acetic acid, the resonance of the carboxyl hydrogen occurs at lower field (by several ppm) than does the hydroxyl proton in the spectrum of water. However, the spectrum of a mixture of the two substances contains only one resonance due to both these kinds of proton. The single resonance observed for the mixture lies between the chemical shifts of the COOH and OH protons in the pure compounds; the shift of the line position from its location in pure water is a linear function of the mole fraction of acetic acid. Hence the measurement of the line position permits the analysis of such a mixture. The observed behaviour arises because the hydroxyl protons of the acetic acid and water are exchanging very rapidly and when the system is examined by NMR spectroscopy we see protons in a time-averaged environment.

In some systems, several kinds of OH, NH or SH groups may be present and the rate of proton exchange between these different environments can be slow compared with the transition time between magnetic energy states. Under these circumstances separate resonances due to each kind of proton will be observed. An increase of temperature, a change in solvent, or the addition of a trace of acid to the sample, may increase the rate of proton exchange and result in the appearance of time-averaged chemical shifts in the spectrum. Occasionally we may chance on a situation in which the mean lifetime before exchange is comparable with the transition time, in which case a broad line is observed. Therefore, OH, NH or SH groups sometimes give rise to characteristically broad resonances. However, broad lines are most frequently observed for NH-protons, because ^{14}N has a nuclear spin and acts like an electric quadrupole. Therefore, depending on the symmetry of the electric field surrounding the ^{14}N nucleus, an attached proton has the opportunity of seeing the ^{14}N nucleus in an average of the several possible magnetic quantum states which are available to it. The broadening of N—H resonances in this manner is said to be due to *quadrupole relaxation*.

If a resonance in an NMR spectrum is suspected to be associated with an OH, NH or SH proton, then the sample solution should be shaken with deuterium oxide and the spectrum repeated. The active hydrogen is replaced by deuterium and the resonance disappears from the spectrum. 'Active' hydrogens which may exchange more slowly with deuterium oxide (such as those attached to nitrogen in amides) can often be replaced if the deuterium oxide contains a trace of a convenient acid, such as trifluoroacetic acid.

Aromatic solvents (e.g., pyridine and benzene) frequently cause chemical shifts of up to 0·5 ppm (relative to CCl_4 or $CDCl_3$), even for C—H protons. Since the various protons in a molecule may be affected to different extents, such shifts may be used in structure elucidation, but a detailed treatment of this topic is beyond the scope of this book. The spectra of very polar samples may be determined in deuterium oxide or dimethylsulphoxide. Tetramethylsilane (TMS) can be retained as the reference material for the latter solvent, but TMS is not miscible with deuterium oxide and hence the sharp methyl resonance of t-butyl alcohol is frequently employed as reference in this case.

C. Hydrogen Bonding

If a proton is hydrogen bonded, then the effect is to cause a downfield shift relative to the unbonded state. The paramagnetic shifts associated with hydrogen bonding may be quite large and it is not uncommon for hydrogen bonded protons of phenols and carboxylic acids to appear at δ values greater than 10 (i.e., at negative τ values). The upfield shift of the hydroxyl proton resonance of ethanol on increasing the temperature, or on diluting the ethanol with carbon tetrachloride, is ascribed to the breaking of intermolecular hydrogen bonds.

Table 3–3 (at the end of the chapter) lists δ-values observed for various types of acidic protons.

3–4. Anisotropy of Carbonyl, Double Bond, and Aromatic Systems

In section 3–3A, we have briefly considered the anisotropy of the carbonyl group; the anisotropy acts to cause deshielding of protons which are held in the plane of the trigonal hybrid bonds (see V). This deshielding will obviously be greatest in the case of an aldehyde proton, directly bonded to the trigonally hybridized carbon atom (see IX). Aldehyde protons do in fact resonate at extremely low field, usually in the $\delta = 9\cdot3$–$10\cdot0$ ppm region.

Olefinic protons, that is those attached to double bonds $\left(\begin{array}{c} H \\ \diagdown \\ C=C \\ \diagup \\ \end{array} \begin{array}{c} H \\ \diagup \\ \diagdown \\ \end{array} \right)$, usually resonate in the $\delta = 4\cdot5$–$6\cdot5$ ppm region. The large deshielding of these protons is attributed mainly to the anisotropy of the π-system, rather than to a simple inductive effect. Experimental measurements of the double bond anisotropy indicate that protons lying near the X- and Z-axes (see X) will be deshielded, while those in the vicinity of the Y-axis will be shielded. Thus, allylic protons $\left(-CH_2-\overset{|}{C}=\overset{|}{C}-\right)$, which must lie near the X- or Z-axes, are deshielded by about 1 ppm relative to methylene protons in a saturated aliphatic chain (see Table 3–1).

IX   X   XI

It is observed that protons attached to systems which can sustain a ring current suffer a paramagnetic shift relative to olefinic protons of isolated double bonds. For example, the aromatic proton frequencies of substituted benzenes usually fall in the 6·5–8·0 ppm region. The reason for this additional deshielding of aromatic protons is that when a benzene nucleus is placed in a uniform magnetic field (H_0), the π-electrons are caused to circulate as indicated in XI. The induced ring current produces a secondary field which reinforces the applied field in the region of the aromatic protons (see XI). Hence the ring current causes the aromatic protons to resonate at a lower value of the applied field. The protons of benzene itself are of course all chemically equivalent and the spectrum of benzene consists therefore of a six proton singlet ($\delta = 7·37$ ppm). Monosubstituted benzenes contain five aromatic protons which can interact extensively and hence they give very complicated spectra; frequently, only partially resolved spectra are obtained. Nevertheless, the effect of monosubstitution on the chemical shifts of the ring protons in benzene has been evaluated, and data are given in Table 3–4 at the end of the chapter.

The values in Table 3–4 illustrate that substituents which are electron-withdrawing (e.g., NO_2—see XIa) cause deshielding of all the aromatic protons, but the deshielding effect is $o > p > m$. How-

XIa   XIb

ever, electron-donating substituents (e.g., NH_2—see XIb) cause shielding of the ring protons; again the effect is frequently $o > p > m$. The paramagnetic shifts for the o- and p-hydrogens of nitrobenzene are larger than the shift of the m-proton, because the mesomeric effect decreases the electron density most at the former positions. An exactly opposite effect operates in aniline (XIb), the electron density being augmented most at the o- and p-positions. The greater shifts of o-hydrogens relative to p-hydrogens may be explained in terms of some combination of steric, anisotropy and inductive effects.

In Tables 3–5A, 3–5B and 3–6, which are appended to this chapter, chemical shifts of protons directly attached to unsaturated linkages are given. The δ values for protons in some simple saturated heterocyclic ring systems are given in Table 3–7 (also at end of chapter).

In subsequent sections, chemical shifts stated without indication of the units refer to ppm on the δ scale.

3–5. Spin–Spin Coupling

In Fig. 3–3† is reproduced the NMR spectrum of 1,1,2-trichloroethane (XII). A doublet centred at 3·95 and a triplet centred at 5·77 are evident in the spectrum. The very small peak at 7·28 is due to traces of chloroform in the $CDCl_3$ solvent, whereas the signal at 3·68 is associated with a trace impurity. In the light of our discussion of inductive effects (section 3–3A), we can understand proton b (adjacent to two chlorine atoms) resonating at lower field than the two a protons (adjacent to one chlorine); but why do the resonances occur as a triplet and a doublet respectively?

$$
\begin{array}{c}
(a) \quad (b) \\
Cl—CH_2—CH
\end{array}
\begin{array}{c}
Cl \\
\diagdown \\
Cl
\end{array}
$$

XII

† Varian catalogue, spectrum no. 2.

The answer lies in spin–spin coupling between the protons attached to adjacent carbon atoms. If we consider the local field experienced by proton *b*, it is evident that the precise value of the field will be dependent on the orientations of the nuclear magnets of the two *a* protons. The nuclear spins of the two *a* protons, which we will designate *a'* and *a''*, can either be (1) both parallel, (2) *a'* parallel and *a''* antiparallel, (3) *a'* antiparallel and *a''* parallel, or (4) both antiparallel. This situation is indicated pictorially in XIII, which takes into account the fact that conditions (2) and (3) are indistinguishable since protons *a'* and *a''* are equivalent. Thus the possible combinations of the nuclear spins of the two *a* protons necessitate that proton *b* will experience three slightly different local fields, and therefore resonate as a triplet. In addition, since opposed spins can occur in two equivalent ways, the intensities of the triplet components will be 1:2:1 (see 5·77 resonance in Fig. 3–3).

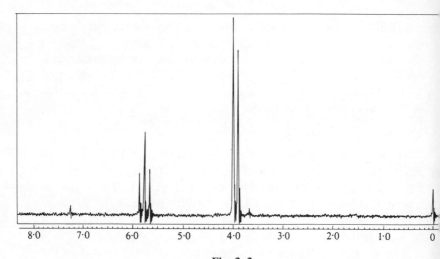

Fig. 3–3

A slightly different way of looking at the same problem uses the approach that since proton *b* can be influenced by two possible orientations of the nuclear magnet of *a'*, the unperturbed signal will initially be split into a doublet. Each component of this doublet is then further split into a second doublet by similar inter-action with *a''*. Since the interactions of *b* with *a'* and *a''* must be equal, the two doublets overlap to give a 1:2:1 triplet (see XIV).

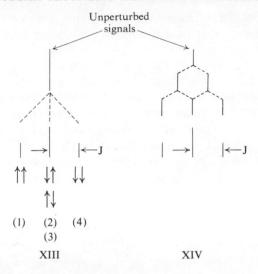

XIII   XIV

The separation of the lines in the triplet gives the coupling constant (J) between the protons b and a' (or b and a''). In this particular case the coupling constant is 6 Hz.

To understand why the a proton resonances occur as a two proton doublet, we must accept that *chemically equivalent protons do not show spin–spin coupling due to interactions among themselves.* Hence the interaction of a' with a'' does not result in observable spin–spin splitting. On the other hand, the chemically equivalent protons a' and a'' are influenced by two possible orientations of the nuclear magnet of b, and therefore their resonance occurs as a two proton doublet (J = 6 Hz).

It is a very simple matter to extend the above reasoning to systems containing additional protons: the following rules can be formulated.

1. If a proton has as neighbours, sets n_a, n_b, n_c, ... of chemically equivalent protons, the multiplicity of its resonance will be $(n_a + 1)(n_b + 1)(n_c + 1)$ .. ...

2. For one neighbouring group of n equivalent protons, the relative intensities of the $n+1$ multiplet components are given by the coefficients of the terms in the expansion of $(x + 1)^n$.

These two important generalizations form the basis for interpreting spin–spin coupling patterns in NMR spectra by the *first-order approximation*. The first-order approximation holds when the chemical shift (δ) between the interacting protons is large compared with the coupling constant (J) between them (in the instance of 1,1,2-trichloroethane (XII), $\delta_{ab} = 109$ Hz and $J_{ab} = 6$ Hz). We will now illustrate the above generalizations by means of two examples, bearing in mind three additional important points. First, large couplings (J = 3–20 Hz) normally only occur between geminal (H—C—H) and vicinal (—CH—CH—) protons. Second, the coupling constant (J) is independent of the applied field, whereas δ_{ab} (in Hz) is proportional to the applied field. Third, proton–proton coupling occurs through the electrons of the bonds between the protons; the proximity of two protons in space is no reason for spin–spin coupling to occur between them.

$$\overset{(a)}{CH_3}—\overset{(c)}{CH_2}—\overset{(b)}{OH}$$

XV

In Fig. 3–4†, the spectrum of ethanol (XV) is reproduced. The protons (a) of the methyl group resonate as a three proton triplet

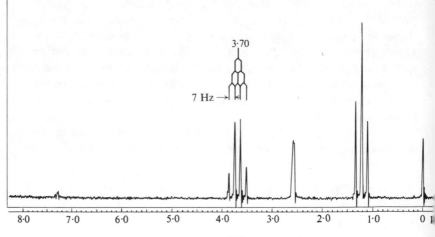

Fig. 3–4

centred at 1·22; a triplet arises because the methyl group has one set of two equivalent protons on an adjacent carbon atom, giving rise to a multiplicity of (2+1), i.e., three lines, whose intensities

† Varian catalogue, spectrum no. 14.

are given by the coefficients of $(x+1)^2$, namely $1:2:1$. The methylene protons (c), being adjacent to the electronegative oxygen atom, resonate at lower field (3·70) in the form of a $1:3:3:1$ quartet. These intensities correspond to the coefficients of $(x+1)^3$ and are consistent with coupling of the methylene protons to one group of three equivalent neighbours. Evidently, no coupling is observed between the methylene protons (c) and the proton (b) of the hydroxyl group. This conclusion is confirmed by the occurrence of the b resonance as a singlet at 2·58. Indeed, not only is the chemical shift of the OH resonance greatly dependent on hydrogen bonding and chemical exchange, but also spin–spin coupling between the protons of a CH—OH system is only observed in the absence of a rapid intermolecular chemical exchange of the hydroxyl protons. Rapid exchange may be prevented either by rigorously freeing the alcohol sample from traces of acids or bases which catalyse exchange reactions, or by employing dimethyl sulphoxide as the solvent for the spectral determination. The final point to note in this spectrum (Fig. 3–4) is that J_{ac} is 7 Hz, a typical value for the coupling constant between vicinal protons in a freely rotating hydrocarbon chain, although the value may be modified slightly by the nature of electronegative atoms directly attached to one of the vicinal carbon atoms (section 3–8A).

In the spectrum (Fig. 3–5)† of 1-nitropropane (XVI), the manner in which the three-proton triplet centred at 1·03 and the two-proton triplet centred at 4·38 (associated with protons a and c respectively) arise will already be evident to the reader. The two equivalent b protons, being coupled to equivalent groups of three and two protons, might be expected to give rise to a two proton resonance with a multiplicity of twelve lines, i.e., $(3+1)(2+1)$ lines. Thus the coupling of the b protons with c protons leads us to expect a triplet, each line of which can be further split into a quartet by the a protons. In practice J_{ab} is approximately equal to J_{bc}, and therefore extensive overlapping of lines occurs as indicated in Fig. 3–5. Since $J_{ab} \simeq J_{bc}$, we can say as a first approximation that the b protons have five equivalent neighbours and that therefore their resonance should be observed as a sextet, the relative intensities of the components being $1:5:10:10:5:1$. This prediction closely corresponds to the actual pattern (Fig. 3–5) centred at 2·07; some distortion of the simple sextet pattern is evident because J_{ab}

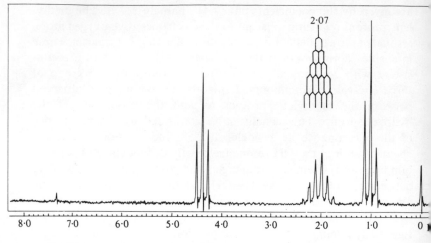

Fig. 3–5

and J_{bc} are not exactly equal. In addition, δ_{ab} is only about ten times as large as J_{ab} and therefore a little perturbation of the line intensities from those predicted by the first-order treatment is apparent. Line perturbations are discussed in more detail in section 3–7.

$$\overset{(a)}{CH_3}\!\!-\!\!\overset{(b)}{CH_2}\!\!-\!\!\overset{(c)}{CH_2}\!\!-\!\!NO_2$$

XVI

3–6. Integration

In the spectra discussed so far in this chapter, it has been possible to estimate the number of protons associated with a particular resonance by 'visual integration' of the spectrum. For more complicated molecules containing many protons, this simple approach cannot be applied with reliable accuracy and the spectrum must be instrumentally integrated.

For a given sample run under a given set of conditions, the total area under all peaks assignable to the protons in a particular environment depends only on the number of such protons and their relaxation times. The latter dependency can be eliminated by operating the spectrometer at very low radio frequency power levels, and hence the areas under the various regions of the spectrum are proportional to the number of protons in the corresponding various chemical environments. The area associated with one proton can usually be determined by integrating the whole spectrum and dividing the whole area by the number of protons in the molecule under investigation. Due to experimental difficulties, an exact integer is seldom obtained, but it is nearly always possible to differentiate between one, two or three protons in a given area. Integration may be carried out immediately after the determination of the spectrum and takes only a few minutes.

The great utility of integration is demonstrated by the spectrum (Fig. 3–6) of oestradiol diacetate (XVII), a steroid containing 28 protons; an integration trace is superimposed on the normal

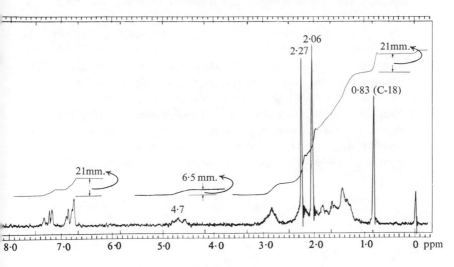

Fig. 3–6

spectrum. It will be noted that this molecule contains an angular methyl group, the protons of which should resonate as a high field three-proton singlet. Indeed, a sharp singlet is evident at 0·83 and since the integration trace rises by 21 mm. on scanning from $\delta = 0·60–0·95$ (a region obviously containing no other resonance signals), one proton corresponds to 7 mm. The 21-mm. interval covering the $\delta = 6·5–7·5$ region is therefore indicative of three aromatic protons, although, before accepting this conclusion, it is wise to confirm that the resonance associated with traces of chloroform in the $CDCl_3$ solvent is negligible in intensity relative to the true aromatic signals. The rise in the integration line of 6·5 mm. on scanning the spectrum from $\delta = 4·2–5·0$ indicates that the broad resonance centred at 4·7 ·is associated with one proton, actually that adjacent to the C—17 acetate group. The splitting patterns associated with these downfield signals cannot profitably be discussed before a detailed consideration of the factors affecting the magnitude of coupling constants (section 3–8).

Other noteworthy features of this spectrum (Fig. 3–6) are the two sharp three-proton resonances at 2·06 and 2·27, associated with the methyl protons of the C—17 and aromatic acetate groups respectively. Each group contains three chemically equivalent protons but no neighbouring vicinal protons; these combined factors account for the sharpness and intensity of the peaks. The paramagnetic shift of the aromatic acetate relative to its aliphatic equivalent is associated with the aromatic ring current (section 3–4).

In contrast, the signals arising from the remaining protons in the molecule are broad and ill-defined ($\delta = 1$–$3\cdot2$ ppm region). The broadness is due to extensive spin–spin coupling between protons of similar chemical shift, which manifests itself in a very large number of unresolved multiplets. It can however be discerned that two of the three benzylic protons (attached to C—6 and C—9) in oestradiol diacetate (XVII) resonate at lower field (2·6–3·1 ppm) than the bulk of the aliphatic protons (1·1–2·5 ppm).

XVII

3–7. Simple Spin–Spin Splitting Patterns

In this section we will designate nonequivalent protons separated by a small chemical shift with the letters A and B, while a third proton separated by a large chemical shift from A and/or B will be indicated by the letter X.

A. AB Systems

An AB system is one consisting of two mutually coupled protons, A and B, which are not coupled to any other protons; in particular, an AB system is one in which J_{AB} is of comparable magnitude to δ_{AB} (the usual notation for $\delta_A - \delta_B$). In contrast an AX system is one in which $\delta_{AX} \gg J_{AX}$.

In outlining the pattern to be expected for an AB system, we will consider the example of a methylene group ($-CH_2-$) with no protons on adjacent carbon atoms. It has been seen in section 3–5 that when such a methylene group is part of an aliphatic chain, the two protons may be chemically equivalent. However, if the methylene group is part of a ring system, or adjacent to an asymmetric centre, or part of an aliphatic chain in which there is restricted rotation, the possibility of nonequivalence of the two protons arises. Under these circumstances, the resonances of the two protons will be split due to mutual coupling of the A and B protons, i.e., the protons will exhibit geminal coupling, usually in the range 12–18 Hz, while the chemical shift between them may be quite small (e.g., 0·1–0·2 ppm). Under these circumstances of small δ/J ratio, the first-order treatment outlined in section 3–5 no longer holds. Thus, instead of four lines all of equal intensity (Fig. 3–7, strictly speaking an AX system), four perturbed lines (Fig. 3–8) are

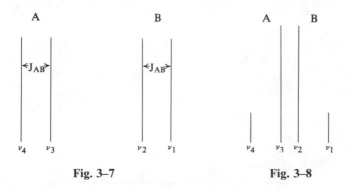

Fig. 3–7   Fig. 3–8

observed. The perturbation occurs so that the bigger peak of the doublet associated with proton B is the lower field signal when A is downfield from B. Conversely, the high field portion of the doublet associated with proton A is more intense than the low field line if B resonates upfield from A (see Fig. 3–8).

The value of the geminal coupling constant is still given in Fig. 3–8 by the separation of the two lines of either of the two doublets.

$$J_{AB} = v_2 - v_1 = v_4 - v_3 \tag{3–8}$$

However, whereas the midpoint between the pair of lines 1 and 2 (or 3 and 4) due to proton B (or A) gives the chemical shift of B (or A) when the δ/J ratio is large (Fig. 3–7), this is not true when the δ/J ratio is small (Fig. 3–8). In the latter case, equation 3–9 holds.

$$\delta_A - \delta_B = \sqrt{(v_4 - v_1)(v_3 - v_2)} \tag{3–9}$$

The relationship which permits calculation of the relative intensity of lines 1 and 2 (or 3 and 4) for a small δ/J ratio is given in equation 3–10.

$$\frac{I_2}{I_1} = \frac{I_3}{I_4} = \frac{v_4 - v_1}{v_3 - v_2} \tag{3–10}$$

To illustrate the actual effect of these relationships, let us consider a hypothetical case in which one proton of an isolated methylene group gives rise to a pair of lines at 48 and 63 Hz whereas the other furnishes lines at 65 and 80 Hz in a 60 MHz spectrum (Fig. 3–9).

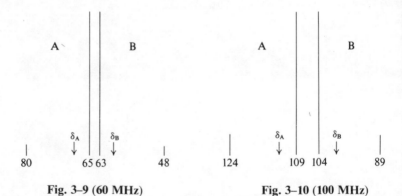

Fig. 3–9 (60 MHz)   Fig. 3–10 (100 MHz)

J_{AB} is evidently 15 Hz, and the relative intensities of the inner and outer pairs of lines is 32/2, i.e., 16:1. A simple calculation, as follows, affords the chemical shifts of A and B.

$$\delta_A - \delta_B = \sqrt{32 \times 2} = 8 \text{ Hz}$$

$$\delta_A = 68 \text{ Hz} \qquad \delta_B = 60 \text{ Hz}$$

The calculation shows that the chemical shifts of A and B lie nearer the strong central lines than the weak satellites. Let us turn now to a 100 MHz analysis of the same signals (Fig. 3–10). Since δ_A and δ_B in Hz are proportional to the operating frequency, the chemical shifts of A and B are now 113·3 and 100 Hz respectively. From equation 3–9, the expression 3–11 can be derived.

$$v_4 - v_2 = v_3 - v_1 = \sqrt{(\delta_A - \delta_B)^2 + J_{AB}^2} \qquad (3\text{–}11)$$

Therefore, since the coupling constant ($J_{AB} = 15$ Hz) is independent of the operating frequency (section 3–5), the line positions in the 100 MHz spectrum are now available to us.

$$v_4 - v_2 = v_3 - v_1 = \sqrt{13\cdot3^2 + 15^2} \simeq 20 \text{ Hz}$$

Employing our knowledge that the centre of the pattern occurs at 106·5 Hz, we can then determine the frequencies as follows:

$$v_4 = 124, \ v_3 = 109, \ v_2 = 104, \ v_1 = 89 \text{ Hz,}$$

and

$$\frac{I_2}{I_1} = \frac{I_3}{I_4} = 7$$

The results are summarized in Fig. 3–10. It can be seen that the separation of the inner pair of lines is increased by a factor of 2·5 when operating at the higher field strength, while the satellites' intensities simultaneously increase more than twofold relative to the central pair. In addition, due to the dependence of signal strength on the square of the frequency, the entire pattern will increase in intensity by a factor of approximately 2·8. Thus, the satellites are approximately six times as intense at 100 MHz as at 60 MHz, illustrating that pattern identification is greatly facilitated by operating at higher frequencies.

The 60 MHz spectrum of columbianetin (XVIII) is given in Fig. 3–11,† and provides an illustration of AB systems.

The one-proton doublet centred at 7·65 is associated with proton i. Generally speaking, protons attached to the β-carbon atom of an $\alpha\beta$-unsaturated carbonyl chromophore resonate at appreciably lower field (6·5–7·8 ppm) than the protons attached to unpolarized double bonds. The polarization in the unsaturated chromophore decreases the electron density at the β-carbon atom (see XIX) and hence results in deshielding of H* (section 3–3A).

† Varian catalogue, spectrum no. 310.

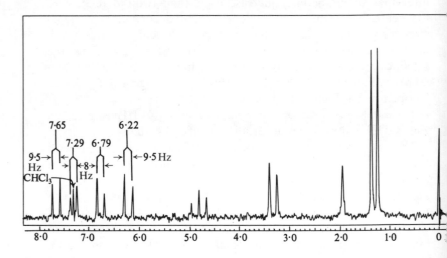

XVIII   XIX

Protons *i* and *f* do in fact form an AB system in which $J_{if} = 9.5$ Hz; the splitting (9·5 Hz) of the one-proton doublet centred at 6·22 tells us that these resonances are associated with proton *f*. The actual chemical shifts of the four lines in this AB system are: 6·13, 6·29, 7·58 and 7·74 ppm, or, in frequency units (from equation 3–6): 368, 377, 455 and 464 Hz. Since the δ/J ratio is not very small (~ 9), the line perturbation should not be very great and can readily be calculated from equation 3–10.

$$\frac{I_2}{I_1} = \frac{I_3}{I_4} = \frac{96}{78} = 1.23$$

This is the order of perturbation observed (Fig. 3–11).

Fig. 3–11

A second AB system in the molecule arises from protons h and g, whose resonances are centred at 6·79 and 7·29 (not necessarily respectively). In this case, $J_{hg} = 8$ Hz, $\delta_{hg} = 30$ Hz and δ/J is smaller (~ 4); consequently, the line perturbation is greater. Also in this spectrum (Fig. 3–11) it should be noted that the two methyl groups, each giving rise to a three-proton singlet (a and b resonating at 1·25 and 1·37, again not necessarily respectively), are adjacent to an asymmetric centre and turn out to be non-equivalent. The hydroxyl proton (c) absorbs as a singlet at 1·96, whereas it transpires that the benzylic protons (d) have the same chemical shift and are split only by coupling to proton e ($J_{de} = 9$ Hz). Proton e (4·82), being coupled equally to both the d protons, resonates as a 1:2:1 triplet.

B. ABX Systems

While the theoretical predictions for the AB pattern are very simple, the situation for the three proton ABX system is slightly more involved. Here it will suffice to outline the pattern as it is most frequently encountered in spectra.

If we combine our treatment of the AB system (section 3–7A) with the simple first order predictions of section 3–5, we may anticipate that the four AB lines will each be split into a doublet, whereas the X proton, being chemically well shifted from A and B and coupled to both, should occur as four lines of approximately equal intensity. Therefore, we may anticipate a 12-line pattern as indicated in Fig. 3–12. This does indeed represent the ABX system

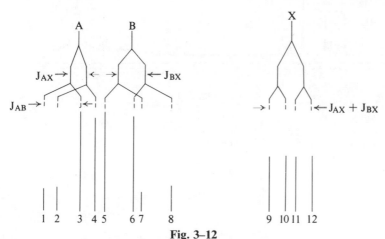

Fig. 3–12

in terms of number of lines as it is usually observed, although the actual manner in which the lines do or do not overlap will of course vary from case to case. However, it is very important to realize that in an ABX system, the observed splittings between lines 9 and 10 and between lines 9 and 11 do not correspond to J_{AX} and J_{BX}, although frequently this may be a good approximation if the chemical shift between A and B is larger than J_{AB}. The relationship which holds for all cases is that the splitting between lines 9 and 12 is equal to the sum of J_{AX} and J_{BX}.

Isolated three-spin systems which can give rise to twelve discernible lines are encountered quite frequently in simple molecules. However, in complicated substances containing many protons, it is quite likely that more than three spins can interact. Nevertheless, the first-order predictions for the ABX system can be utilized in a wide variety of cases. For example, in the spectrum (Fig. 3–13)† of 1-cyclopentylbut-1-en-3-one (XX) the olefinic protons c and d,

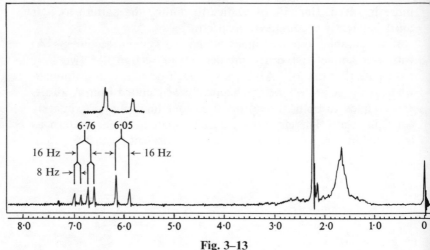

Fig. 3–13

XX

† Varian catalogue, spectrum no. 544.

whose resonances are centred at 6·05 and 6·76 respectively, can be regarded as the A and B protons of an ABX system, the cyclopentyl proton b being designated X. This is permissible since the protons whose signals we wish to analyse (A and B) are coupled only to X, although the X proton is of course further coupled to other protons in the cyclopentyl ring.

The resonance of proton d (at 6·76) is split into a quartet with $J_{cd} = 16$ Hz and $J_{bd} = 8$ Hz. The line perturbation is associated with the larger coupling (J_{cd}), indicating that proton c lies close at hand. The c resonance (6·05) appears at first sight to be a doublet, but on spreading the signals out more (see superimposed trace in Fig. 3–13), four lines can just be resolved. The very small splitting corresponds to J_{bc} (1 Hz); it will be seen in section 3–9 that small spin–spin couplings (0–3 Hz) between protons separated by four bonds is not uncommon, especially when one of the intervening bonds involves a π-linkage. The X proton (b) in this spectrum (Fig. 3–13) of course absorbs at high field (probably about 2·6 ppm). The sharp peak at 2·23 is associated with the protons (a) of the methyl group.

3–8. Factors Affecting Coupling Constants

The reader may have noted that in the previous example we observed a wide range of coupling constants. The factors which determine the magnitude of coupling constants have been widely studied in the last few years, and the most important ones will be indicated in this section.

A. Vicinal Protons (H—C—C—H)

(i) Dihedral angle dependence. The relationship between the vicinal coupling constant (J) and the dihedral angle (ϕ) between the protons (see Fig. 3–14 for the definition of ϕ in this context) is given approximately by the theoretically derived Karplus equation (equation 3–12).

$$J = \begin{cases} 8\cdot5 \cos^2 \phi - 0\cdot28, & 0° \leqslant \phi \leqslant 90° \\ 9\cdot5 \cos^2 \phi - 0\cdot28, & 90° \leqslant \phi \leqslant 180° \end{cases} \quad (3\text{–}12)$$

It will be seen in the subsequent parts of this section that many additional variables are important in deciding the magnitude of J_{vic}; so we may profitably forget about the small constant (0·28), and get a rough pictorial idea of what the Karplus equation implies from the plot of J against ϕ (Fig. 3–14).

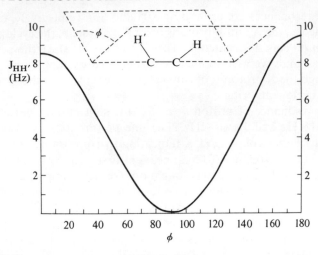

Fig. 3–14

It is apparent from the figure that J will be largest when the vicinal protons are trans-coplanar, slightly smaller when they are cis-coplanar, and around zero when the protons are at right angles. One of the most important consequences of the Karplus equation is that the order of magnitude of diaxial, axial-equatorial and diequatorial coupling constants (J_{aa}, J_{ae} and J_{ee} respectively) in a cyclohexane ring chair system (**XXI**) can be predicted and is in reasonable agreement with the observed values (see Table 3–8).

Table 3–8
Coupling Constants (Hz) in Six-Membered Ring Systems

	Karplus equation	Observed	
		Usual	Widest range
$\phi_{aa} = 180°$ J_{aa}	9	10–13	8–14
$\phi_{ae} = 60°$ J_{ae}	1·8	2–5	1–7
$\phi_{ee} = 60°$ J_{ee}	1·8	2–5	1–7

The wide range quoted in the table includes a number of exceptionally large and small experimental values, whereas the 'usual' column covers about 90 per cent of the cases which will be encountered. The important conclusion to be derived from Table 3–8 is that large vicinal coupling constants (10–13 Hz) may be identified with an approximate diaxial orientation of the atoms (see XXII, which is a projection of the cyclohexane ring system XXI viewed along the 1—2 bond), while smaller splittings (2–5 Hz) are associated with axial-equatorial or diequatorial interactions.

XXI   XXII

A modified Karplus equation can also be applied to vicinal couplings in olefins (H—C=C—H') and leads to the prediction that *trans*-olefinic coupling constants ($\phi = 180°$) will be larger than *cis*-olefinic coupling constants ($\phi = 0°$). This prediction is borne out in practice, as can be seen from the quoted experimental ranges which are observed for acyclic systems.

$J_{H,H'}(cis) = 7–11$ Hz $J_{H,H'}(trans) = 12–18$ Hz

(*ii*) *Dependency on electronegativity of substituents.* For a freely rotating alkyl chain (e.g., XXIII) the dependence of $J_{H,H'}$ upon the electronegativity of X is given by equation 3–13.

$$J = 7·9 - 0·7 \, \Delta E \qquad (3–13)$$

In equation 3–13, 7·9 is close to the observed coupling constant of ethane ($J_{H,H'} = 8·0 \pm 0·2$ Hz)† and ΔE is equal to ($E_X - E_H$), where E_X and E_H are the Huggins electronegativities of the substituent and hydrogen respectively. The equation only gives an approximate

† It has been pointed out previously that chemically equivalent protons do not show spin–spin coupling due to interaction among themselves. In these cases, the coupling constants may be evaluated by isotopic substitution.

coupling constant for di- or tri-substitution by electronegative atoms. However, it gives a useful guide to the coupling constant (6 Hz) between vicinal protons for 1,1,2-trichloroethane (XII, see Fig. 3–3). The Huggins electronegativities for hydrogen and chlorine are 2·20 and 3·15 respectively; the anticipated vicinal coupling constant is therefore $[7·9 - (3 \times 0·95 \times 0·7)]$, i.e., 5·9 Hz, in reasonable agreement with the observed value.

$$CH_3—CH_2'—X \qquad\qquad Cl—CH_2—CH—Cl_2$$
$$\text{XXIII} \qquad\qquad\qquad \text{XII}$$

For cyclic systems, it has been established that the effect of an electronegative substituent (X) on the vicinal coupling constant varies with the orientation of X with respect to the coupling protons. Thus, J_{ae} in a cyclohexane chair system is about 5·5 Hz (± 1 Hz) for an equatorial substituent (XXIV, X = OH, OAc, Br), but about 2·5 Hz (± 1 Hz) for an axial substituent (XXV, X = OH, OAc, Br). These differences arise despite the similar dihedral angles ($\sim 60°$) involved in both cases. We can already see why relatively large deviations from the values predicted by the Karplus equation are observed. Other electronegative groups give rise to similar effects.

XXIV   XXV

In XXIV and XXV the bonds through which the relevant vicinal couplings occur are marked in heavy outline. Apparently, the effect of X in reducing J_{vic} is greater when X is *trans*-coplanar with respect to the C—2 proton (H_a in XXV) than when the angle between X and the C—2 proton (H_e in XXIV) is about 60°. Hence the configuration (axial or equatorial) of electronegative substituents may be deduced from the magnitude of the associated axial-equatorial coupling constants.

The spectrum (Fig. 3–15)† of 3,4-epoxytetrahydrofuran (XXVI)

---

† Varian catalogue, spectrum no. 405.

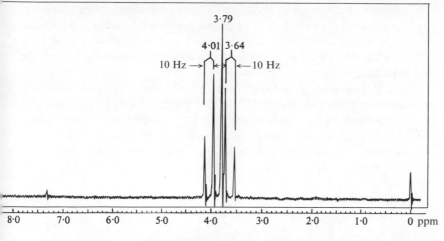

Fig. 3–15

illustrates how the knowledge that vicinal coupling constants are dependent upon the orientation of electronegative substituents may greatly facilitate the interpretation of a spectrum.

XXVI

3,4-Epoxytetrahydrofuran (XXVI) has a plane of symmetry and therefore gives rise to only three distinct proton resonances. A little surprisingly, the two b protons resonate as a two-proton singlet at 3·79; the relative chemical shifts of a, b, and c can be varied by determining the spectrum in a wide variety of solvents, but the splitting pattern is unchanged. From the conformation in which the compound is held (see XXVI) it is not surprising that no appreciable coupling takes place between b and c, since the dihedral angle between them is around 90°. However, an analysis based only on a knowledge of the Karplus equation would require an appreciable coupling ($\sim$6–8 Hz) between protons b and a (protons approximately eclipsed). No significant coupling is observed, because the electronegativity effect of the epoxide is transferred more efficiently to a than c. As a consequence of this behaviour, a and c are coupled only with each other; each resonates as a doublet (J_{ac} = 10 Hz). The doublets are centred at 3·64 and 4·01, but cannot specifically be assigned to a or c.

(*iii*) *Dependency on* C—C—H *bond angles.* Vicinal coupling constants depend markedly on the angles θ and θ' (see **XXVII**) subtended by the carbon–carbon and carbon–hydrogen bonds. This effect is most easily demonstrated by reference to cyclic olefins, because the *cis*-olefinic coupling constants ($J_{H,H'}$—see **XXVIII–XXXI**) preclude a variation in the dihedral angle (ϕ = 0°) and $\theta = \theta'$.

XXVII   XXVIII, J = 0·5–2·0 Hz   XXIX, J = 2·5–4·0 Hz

Obviously, as the ring size is decreased from a six-membered ring (a cyclohexene, **XXXI**) to a three-membered ring (a cyclopropene, **XXVIII**) through intermediate five-(**XXX**) and four-(**XXIX**) membered systems, the angles (θ and θ') between the

XXX, J = 5·1–7·0 Hz   XXXI, J = 8·8–10·5 Hz

olefinic protons will be increased. These increases are reflected in successive decreases in J_{vic} as the ring size is reduced (see formulae). The quoted range of J given along with each structural formula is based on data for several compounds containing that particular size of ring.

Vicinal coupling constants observed in some carbocyclic rings (**XXXII**) and cyclic ethers (**XXXIII**) are given in Table 3–9. The observed values within the quoted ranges are dependent upon the electronegativity of any substituents. However, as a general rule, in monosubstituted cyclopropanes and epoxides, J_{trans} is less than J_{cis}, as illustrated for cyclopropanecarboxylic acid (**XXXIV**) and styrene

Table 3–9

Vicinal Coupling Constants (Hz) in Cyclic Systems

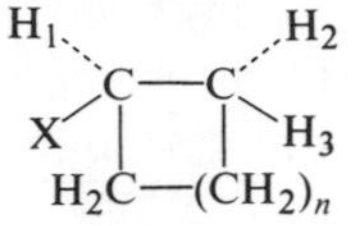

XXXII   XXXIII

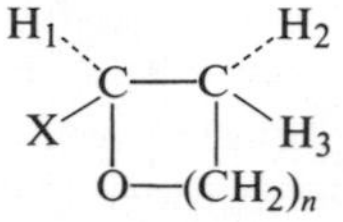

n	$J_{cis}(J_{1,2})$	$J_{trans}(J_{1,3})$	$J_{cis}(J_{1,2})$	$J_{trans}(J_{1,3})$
0	8–13	5–10	2–5	1–2·5
1	5–11	2–11		
2	1–9	1–9	1–9	1–9
3	2–5†	10–13‡ or 2–5§	—	

† axial-equatorial.   ‡ diaxial.   § diequatorial.

oxide (**XXXV**); this relationship is consistent with the dihedral angles involved, and often holds for cyclobutanes too.

Insufficient data are available to give generalizations for four-membered rings containing an oxygen atom. In cyclopentane rings, *cis*- and *trans*-coupling constants are very similar, and the stereo-chemical relationship (*cis* or *trans*) of 1,2-substituents cannot readily be evaluated from coupling constant data. The similarity arises because, in a cyclopentane ring (see **XXXVI**), *cis*-protons are almost eclipsed and hence J_{cis} is almost optimal, while *trans*-protons cannot become coplanar (dihedral angle 180°) without

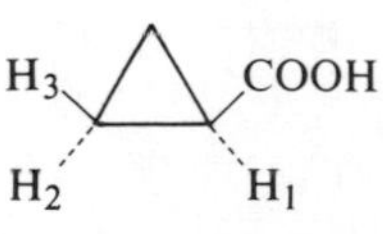

XXXIV, $J_{1,2} = 8{\cdot}0$ Hz   XXXV, $J_{1,2} = 4{\cdot}0$ Hz   XXXVI
$J_{1,3} = 4{\cdot}6$ Hz $J_{1,3} = 2{\cdot}5$ Hz

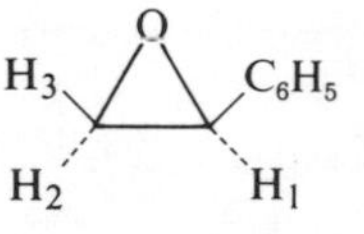

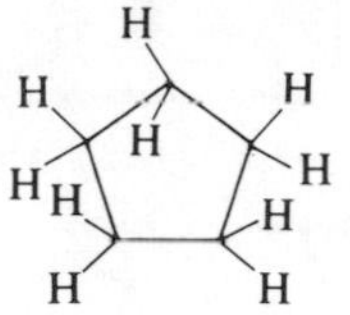

distorting the ring. Since four- and five-membered rings can undergo subtle conformational changes depending on the substituents, a wide range of J values is feasible. The most important generaliza-tions are that: very large coupling constants (>11 Hz) are not observed, and values most commonly lie in the range 5–9 Hz. The values for cyclohexane rings (XXXII, $n = 3$) are included in the table for comparison purposes.

(*iv*) *Bond length dependence.* For constant bond angles and hybridization, the vicinal coupling constant decreases with increasing C—C bond length. This factor is most important in considerations of unsaturated and aromatic systems, in which variations of π-bond order occur. *Cis*-olefinic coupling constants for a number of compounds are indicated in formulae XXXVII–XL; the figures in parentheses quoted after coupling constants refer to the π-bond orders of the intervening C—C bonds.

It is apparent that the *cis*-olefinic coupling constant is appreciably greater in ethylene (XXXVII, short C—C bond, π-bond order 1) than in benzene (XXXVIII, longer C—C bond, π-bond order 0·67). Similarly in naphthalene (XXXIX), $J_{1,2}$ is greater than $J_{2,3}$ because the 1,2-bond has the greater π-bond order. In pyridine (XL), $J_{2,3}$ is less than $J_{3,4}$, in sharp contrast to the relative magnitudes anticipated on the basis of π-bond order alone. It is apparent that the reduction in $J_{2,3}$ caused by the electronegative nitrogen atom overrides the π-bond order effect.

Effect of π-bond Order (Bond Length) on Cis-olefinic Coupling Constants (Hz)

XXXVII   XXXVIII

XXXIX   XL

The utility of the characteristic variations of *cis*-olefinic coupling constants in structure elucidation may be illustrated by the spectrum (Fig. 3–16)† of XLI.

<hr>

† Varian catalogue, spectrum no. 327.

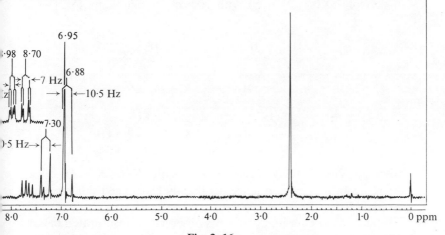

XLI

The most obvious features of the spectrum are the three-proton singlet at 2·45, and the two-proton singlet at 6·95, corresponding to protons *a* and *d* respectively. Protons *b* and *c* give rise to an AB system; the resonances are centred at 6·88 and 7·30 (not necessarily respectively) and the large *cis*-olefinic coupling constant (10·5 Hz) indicates that the double bond is of the ethylenic type, rather than the benzenoid. Proton *e* resonates as a quartet ($J_{2,3}$ = 5 Hz and $J_{3,4}$ = 7 Hz) centred at 7·68 which is compatible with its location at C—3 of a pyridine ring with neighbouring protons attached to C—2 and C—4. The resonances of *f* and *g* at 8·70 and 8·98 show the expected splittings of 7 Hz and 5 Hz respectively; each doublet which can be regarded as arising in this manner is then further split into a quartet by a meta-coupling ($J_{2,4}$) of 2 Hz. It will be seen in section 3–9A that meta couplings of this order of magnitude are usual.

Fig. 3–16

A combination of all the factors discussed in this section must be considered to enable us to predict the *cis*-olefinic coupling constants in the five-membered heterocyclic compounds furan (XLII) pyrrole (XLIII) and thiophene (XLIV). The C—C bond lengths in these compounds are all very similar, but the relevant bond angles show a slight increase in the series thiophene→ pyrrole→furan (see formulae). The electronegativities of the heteroatoms are in the order $O > N > S$. Therefore it must be concluded that $J_{2,3}$ will be greatest in thiophene (XLIV), intermediate in pyrrole (XLIII) and least in furan (XLII), as confirmed by experiment (see formulae; the quoted ranges represent values for several substituted heterocycles).

XLII

$J_{2,3} = 1\text{–}2$ Hz

XLIII

$J_{2,3} = 2\text{–}3$ Hz

If it is assumed that the angles in thiazole (XLV, X = S) and oxazole (XLV, X = O) are similar to those quoted for thiophene (XLIV) and furan (XLII) respectively, then because of the electronegativity of the nitrogen atom, $J_{2,3}$ should be algebraically decreased in thiazole (XLV, X = S) with respect to thiophene (XLIV) as

XLIV

$J_{2,3} = 4\cdot5\text{–}5\cdot5$ Hz

XLV

$X = S, J_{2,3} = 3\cdot4$ Hz

observed. The 60 MHz spectrum of oxazole (XLV, X = O) shows the H—2 and H—3 resonances as slightly broadened singlets, indicating that $J_{2,3}$ must be very small.

Some vicinal coupling constants observed in unsaturated heterocyclic systems are summarized in Table 3–10 at the end of the chapter.

B. Geminal Protons (H—C—H)

The sign of the geminal coupling constant (J_{gem}) is usually negative and opposite to that of the vicinal coupling constant which is nearly always positive. This change in sign need not perplex the chemist who wishes to interpret his spectra as far as possible by the first-order approximation; it is important only in evaluating the direction of change in J_{gem} which is produced by various factors.

(*i*) *Dependency on electronegativity of substituents.* The geminal coupling constant between the methylene protons of a system RCH_2X (XLVI) increases algebraically with increasing electronegativity of X. The effect is illustrated below for methane, methanol and methyl fluoride and, although not large, may be useful in structural work. Geminal coupling constants between the equivalent protons of a methyl group cannot of course be evaluated directly. A useful technique is to replace one of the equivalent protons by deuterium; H—D coupling may then be observed and the proton–proton coupling constant calculated by means of equation 3–14.

$$J_{H,H} = 6.55J_{H,D} \qquad (3\text{–}14)$$

In contrast, J_{gem} decreases algebraically with increasing electronegativity of a substituent (X) attached to the carbon atom adjacent to the geminal protons in a series of monosubstituted 1,1-dichloro-cyclopropanes (XLVII). The opposite effects of electronegativity of X upon J_{gem} in systems XLVI and XLVII forcibly illustrates that changes in J are not simply attributable to the direct inductive effect of the substituent.

$$R—CH_2—X$$

Cl   Cl

H   X

H   II

XLVI, R = X = H, $J_{gem} = -12.4$ Hz
   R = H, X = OH, $J_{gem} = -10.8$ Hz
   R = H, X = F, $J_{gem} = -9.6$ Hz

XLVII, X = Si(CH$_3$)$_3$, $J_{gem} = -4.9$ Hz
   X = COOH, $J_{gem} = -6.8$ Hz
   X = OAc, $J_{gem} = -9.7$ Hz

(*ii*) *Angle dependence.* Geminal coupling constants undergo a marked algebraic increase on increasing the angle θ (see XLVIII) between the interacting protons. Data for some cyclic systems are summarized in Table 3–11.

$$\underset{\text{XLVIII}}{\overset{\displaystyle H \underset{\theta}{\diagdown} \underset{C}{} \diagup H}{\underset{\theta'}{\diagdown}}} \qquad \underset{\text{XLIX}}{\overset{\displaystyle H \underset{120°}{} H}{}}$$

In the relatively strain-free cyclohexane and cyclopentane rings (L, $n = 4$ and 3 respectively), the angle θ is similar to the tetrahedral angle ($\theta = 109°$) and J_{gem} is accordingly reminiscent of the value for methane. On decreasing the ring size to a cyclopropane system (L, $n = 1$), θ' is decreased with consequent increase in θ (see XLVIII). The limiting case is reached when $n = 0$, and instead of a cyclic system we have a terminal methylene group in which the carbon atoms are sp^2 hybridized and $\theta = 120°$ (XLIX); the geminal coupling may then even become positive. The range quoted for cyclopropanes (L, $n = 1$) corresponds to a wide selection of electronegative substituents attached to C—2.

The coupling constants between geminal protons adjacent to oxygen in cyclic ethers (LI) show analogous variations with ring size. Most noteworthy is the change of J_{gem} from negative to positive

Table 3–11

Geminal Coupling Constants (Hz) in Cyclic Systems

n	L — J_{gem}	LI — J_{gem}	LII — J_{gem}
0	0–3†	—	—
1	−(4–9)	4–7	—
2	−(7–14)‡	—	0–2†
3	−(10–14) - - - - - - - -(8–14)		−(5–7)
4	−(12–14) - - - - - - -(10–14)		—

† Sign undetermined in most cases
‡ Usually −(10–14); numerically smaller values in strained systems

on passing from cyclopropanes (L, $n = 1$) to epoxides (LI, $n = 1$; the quoted range includes substituted epoxides); the change is anomalously large and may be attributed to the unique hybridization of three-membered rings.

Finally, methylenedioxy groups in five- (LII, $n = 2$) or six- (LII, $n = 3$) membered rings exhibit the anticipated J values. The effect of introducing one and two oxygen atoms successively into five- and six-membered rings is nicely illustrated by comparing the values joined by dotted lines in the table.

(*iii*) *π-contribution*. The magnitude of geminal coupling constants is greatly dependent on the number of π-bonds adjacent to the methylene or methyl group. If the geminal protons can assume all possible rotational conformations with respect to the adjacent π-system, then it is found that the negative geminal coupling constant is decreased algebraically (i.e., becomes more negative) by about 1·9 Hz for each adjacent π-bond. This situation is indicated graphically in Fig. 3–17, deviations from linearity being less than 1 Hz. Actual values are given for methane, toluene, acetonitrile and malononitrile; the relevant number of π-bonds is given in parentheses after the coupling constant in each case.

In cyclic systems, methylene protons may have a fixed orientation relative to an adjacent π-system. Theoretical approaches in fact

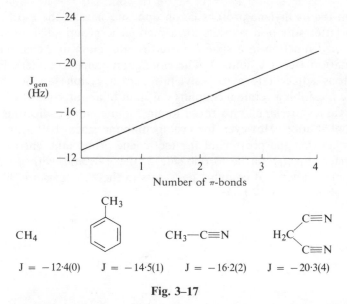

Fig. 3–17

predict that the π-contribution to geminal coupling (J^π) will vary with the angle (ϕ) subtended between the π-orbital and one of the methylene protons (see LIV which is a view of LIII along the bond joining the sp^2 and sp^3 hybridized carbon atoms). The variation of J^π with ϕ is given in Fig. 3–18.

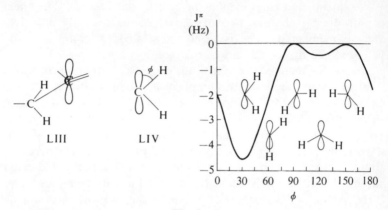

Fig. 3–18

It is apparent from Fig. 3–18 that the π-contribution to geminal coupling will be largest when both the methylene protons lie to one side of the π-orbital ($\phi = 10°–50°$). In contrast, J^π will be small when the methylene protons lie on opposite sides of the π-orbital. The latter situation prevails for a methylene group adjacent to a carbonyl function in a six-membered ring existing in a chair conformation (see LV and a). Therefore, non-equivalent methylene protons adjacent to a carbonyl group in a cyclohexanone chair system exhibit a geminal coupling constant of about $-(12–13$ Hz); the values are similar to those found for methane, indicating no π-contribution. However, for cyclopentanone rings (LVI), models indicate that the protons of the methylene group adjacent to the π-system must both lie to one side of the π-orbital, with $\phi \simeq 20°$ (see b); the geminal coupling constants in these systems are therefore about $-(16–17$ Hz).

In 2-methyl-4-oxacyclopentanone (LVII), the nonequivalent d and e protons are those of an isolated methylene group and therefore show up in the spectrum (Fig. 3–19)† as an AB system; the

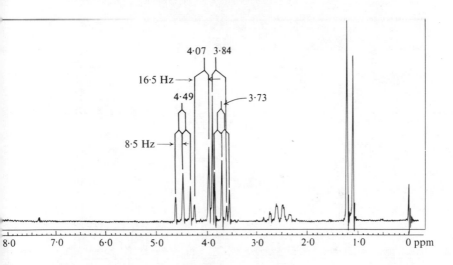

LVII

resonances are centred at 3·84 and 4·07 ppm. The very large geminal coupling constant ($-16·5$ Hz) suggests (but certainly does not prove) their presence in a five-membered ring adjacent to a π-system. The c and f methylene protons are also non-equivalent and, in addition to being mutually coupled, can interact with

Fig. 3–19

proton b. It turns out that $J_{cf} = J_{bc} = J_{bf} = |8·5$ Hz$|$, which is not too surprising from the data accumulated in Table 3–9 and Table 3–11. Hence protons c and f resonate as triplets centred at 3·73 and 4·49 respectively, as indicated in the figure.

† Varian catalogue, spectrum no. 438.

J_{ab} (7 Hz) is similar, but not identical to J_{cf}, J_{bc} and J_{bf}. It follows as a first approximation therefore that proton b (2·55 ppm) undergoes five similar interactions with vicinal protons and its resonance consequently occurs roughly as a 1:5:10:10:5:1 sextet. The typical secondary methyl group resonance is apparent at 1·17 ppm.

In the above example, the difference between the geminal coupling constants $(J_{de} - J_{cf})$ is a measure of the π-contribution $(J^{\pi} = -8$ Hz). This value is appreciably bigger than the maximum theoretical value $(J^{\pi} = -4·5$ Hz), but represents the maximum deviation which is likely to be encountered. The spectrum (Fig. 3–20)† of the isomeric 3-methyl-4-oxacyclopentanone (LVIII)

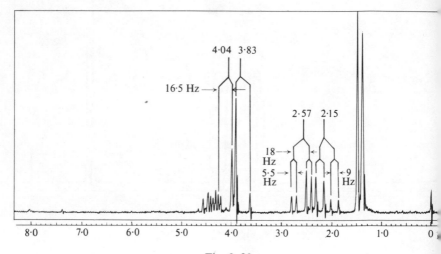

LVIII

Fig. 3–20

again shows an AB system associated with protons d and e $(J_{d,e} = 16·5$ Hz). In this case, protons b and c also constitute a geminal pair adjacent to carbonyl and afford the eight lines of the

† Varian catalogue, spectrum no. 439.

AB part of an ABX system; $J_{bc} = 18$ Hz, $J_{bf} = 5.5$ Hz and $J_{cf} = 9$ Hz. The resonance of proton f, centred at 4·35, is rather complicated, because the proton is coupled to no less than five vicinal neighbours. However, if one knows the coupling constants involved, it is a simple matter to derive the observed pattern. Since $J_{af} = 6$ Hz (see doublet methyl resonance), the unperturbed resonance of f initially is split by four almost equal interactions; namely, $J_{af} = 6$ Hz ($\times 3$) and $J_{bf} = 5.5$ Hz ($\times 1$), which should lead as illustrated in Fig. 3–21 to an approximate $1:4:6:4:1$ quintet. Each line of this quintet is then split into a doublet because of the spin-coupling of f to c ($J_{cf} = 9$ Hz). The result is a ten-line pattern (relative intensities $1:4:1:6:4:4:6:1:4:1$, see Fig. 3–21),

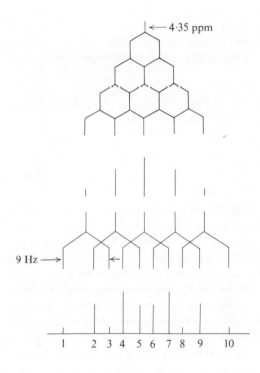

Fig. 3–21

which is almost exactly reproduced in Fig. 3–20, although line 8 is actually hidden beneath one of the lines associated with the AB system due to protons d and e.

3–9. Long-Range Spin–Spin Coupling

So far we have concerned ourselves only with spin couplings occurring between protons attached either to the same carbon atom or to adjacent carbon atoms; these cases represent coupling through two and three bonds respectively. In practice, appreciable couplings (0–3 Hz) are frequently observed between protons separated by four or five bonds. These interactions we shall define as long-range couplings; they are particularly common in π-systems, and a knowledge of their magnitude and stereochemical requirements can be of great assistance in the interpretation of an NMR spectrum.

A. Substituted Benzenes

Long-range couplings are important in aromatic compounds. The ranges of coupling constants which are observed in substituted benzenes are summarized in LIX; the coupling constants are attenuated with increasing number of intervening bonds.

$J_{AO}(ortho) = 6\text{–}10$ Hz
$J_{AM}(meta) = 1\text{–}3$ Hz
$J_{AP}(para) = 0\text{–}1$ Hz

LIX

LX

The spectra of many *para*-disubstituted benzenes (see LX) approximate to A_2X_2 types. In general, A_2X_2 spectra fall into two types according to whether there is one AX coupling constant or two. In 1,1-difluoroallene (LXI) there is only one AX coupling constant,† but in *p*-disubstituted benzenes such as *p*-bromophenetole (LXII) there are two AX coupling constants. In the latter case, the first-order approximation breaks down; it is not true to say spectra of the A_2X_2 type are independent of J_{AA} and J_{XX} when $J_{AX} \neq J'_{AX}$. Quantum mechanical considerations lead to the prediction of no less than ten lines for each nucleus (A and X).

In practice, the spectra of *p*-substituted benzenes usually show quite clearly a few of the additional lines predicted by theory. If the system was naïvely treated as two identical AB systems (LX, with the symbol 'X' replaced by 'B') which could not interact,

† As might be anticipated from the first order approximation, these systems give rise to symmetrical (1:2:1) triplets for A and X with spacings equal to J_{AX}.

four lines (two perturbed doublets) with a splitting of each doublet equal to J_{AB} would be anticipated. At a first glance at the spectrum (Fig. 3–22)† of *p*-bromophenetole (LXII), this seems to be the case.

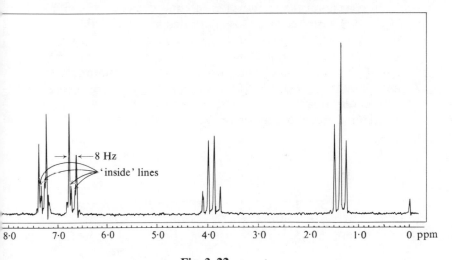

LXI   LXII

In actual fact, the splitting of 8 Hz does not give J_{AX} (see **LXII**), but turns out to be a good approximation to the coupling constant in these cases. The most important point for analytical work is that the 'quartet' in the aromatic region with additional weak lines

Fig. 3–22

surrounding the four main signals is some evidence *a priori* for a *p*-disubstituted benzene or 4-substituted pyridine. In particular, the four extra 'inside lines' (see Fig. 3–22) are quite characteristic.

† Varian catalogue, spectrum no. 198.

B. Allylic Coupling

Allylic coupling is defined as spin–spin coupling occurring between H_1 and H_3 or H_2 and H_3 in the unsaturated system LXIII; $J_{1,3}$ is described as a *transoid*-allylic coupling constant, whereas $J_{2,3}$ is a *cisoid*-allylic coupling constant. Allylic coupling constants normally lie in the range 0–3 Hz, but their variations with interbond angles are quite difficult to predict.

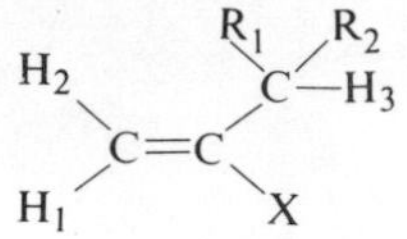

LXIII

The spectrum (Fig. 3–23)† of plumericin (LXIV) may be used to demonstrate the effects of allylic coupling. Assigning first the most obvious signals, the three-proton singlet at 3·79 is due to the hydrogens (c) of the methoxyl group and the three-proton doublet ($J_{ai} = 7$ Hz) centred at 2·11 is associated with the methyl group (a) of the ethylidene moiety. Proton f, adjacent to two oxygen atoms, should resonate at quite low field as a doublet since it can only interact with a single vicinal neighbour. The appropriate signals are centred at 5·59 ($J_{bf} = 6$ Hz); the coupling constant is consistent with a very small dihedral angle between b and f, as indicated by a Dreiding model of plumericin (see LXV).

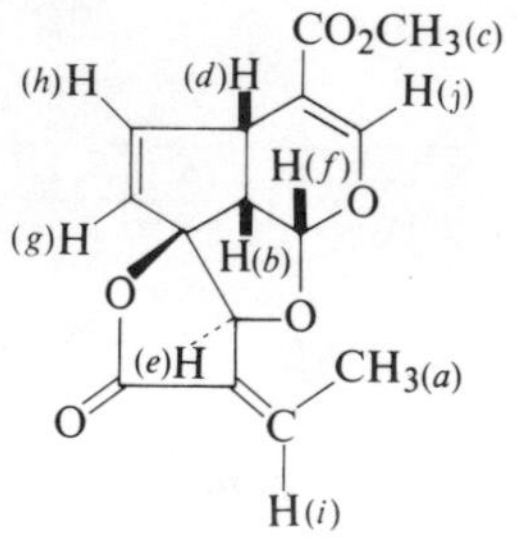

LXIV 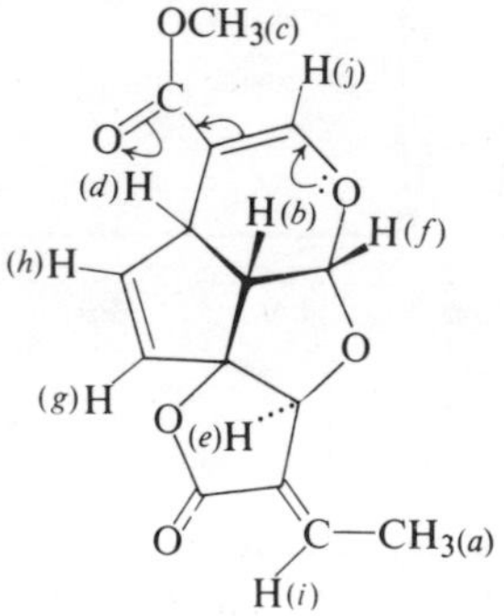   LXV

† Varian catalogue, spectrum no. 640.

Proton b should be evident therefore as a four-line pattern with one splitting of 6 Hz (J_{bf}); the signals are centred at 3·46 and since b and d are almost perfectly eclipsed (see LXV), J_{bd} is large (9 Hz). The olefinic hydrogen i is allylically coupled to proton e, and therefore the 1:3:3:1 quartet centred at 7·19 and produced by coupling of i with the a protons is further split into a total of eight lines (J_{ai} = 7 Hz, J_{ei} = 1·5 Hz). Similarly, protons d and g are suitably

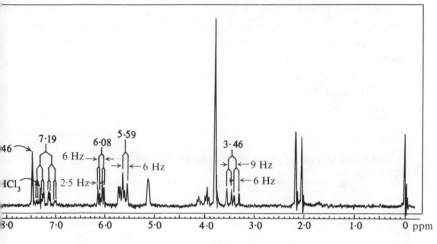

Fig. 3–23

placed for allylic coupling to occur between them. Indeed, J_{dg} and J_{dh} are both 2·5 Hz, and so the quartets occurring at 5·67 and 6·08 can be assigned to the g and h protons *but not specifically to one or the other*. It should be noted that J_{gh} (6 Hz) is typical for the coupling between olefinic hydrogens in a five-membered ring.

In contrast, proton j (singlet at 7·46) does not appear to be significantly coupled to d. Apparently, the conformation of the six-membered ring is such that the spatial relationship between the protons is unsuitable for observable allylic coupling. Also, it is possible that the π-bond order of the double bond may be decreased by the type of delocalization indicated in LXV, with consequent decrease in the π-contribution to J_{al} ($J_{allylic}$). Such resonance can account for the difference in allylic coupling constants in LXVI and LXVII. Finally, the signal (5·12) associated with proton e is obviously broadened by allylic coupling, even though two distinct lines are not apparent.

For plumericin (LXIV) it was noted that J_{dh} and J_{dg} were identical (2·5 Hz). $J_{1,3}$ and $J_{2,3}$ values for the parent cycloalkenes are

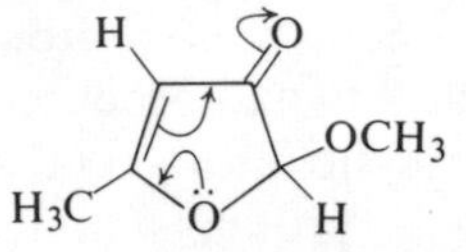

LXVI, J_{al} = 0·8 Hz 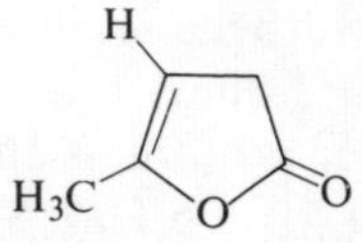   LXVII, J_{al} = 1·6 Hz

recorded in Table 3–12. The variations of these coupling constants with ring size are of some structural utility, but it must be remembered that six-membered ($n = 3$) and seven-membered ($n = 4$) rings are conformationally mobile and therefore $J_{2,3}$ and $J_{1,3}$ depend greatly on the type of structure into which the ring is incorporated.

Table 3–12

Comparison of Vicinal ($J_{2,3}$) and Allylic ($J_{1,3}$) Coupling Constants (Hz) in Cycloalkenes

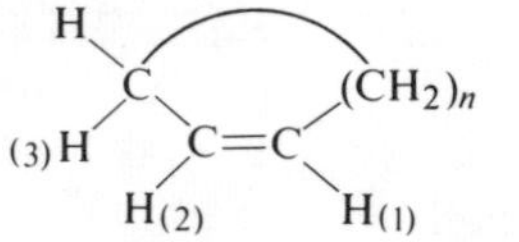

	n	$J_{2,3}$	$J_{1,3}$
	0	±1·8	—
	1	−0·8	+1·5
	2	2·1	−1·4
	3	3·1	−1·4
	4	5·7	−1·0

C. Homoallylic Coupling

Coupling between H(1) and H(4) in LXVIII is described as homoallylic coupling. As might be expected, homoallylic coupling constants are very small (0–2·0 Hz). The interaction is greatest when both θ and θ' (see LXIX) are around 90° and usually negligible if these angles are small ($<25°$). The coupling can occur when X and Y are *cis-* or *trans*-substituents of the double bond.

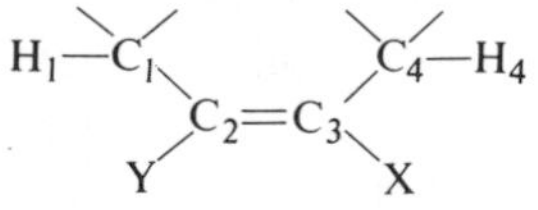

LXVIII 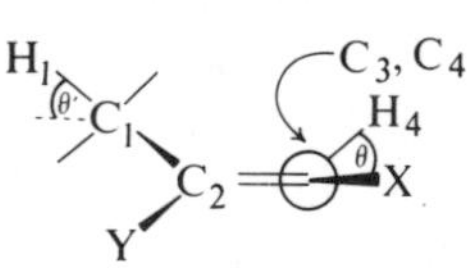   LXIX

Obviously the stereochemical requirements of homoallylic coupling are more likely to be satisfied if one of the groups attached to the double bond is freely rotating (e.g., a methyl group), than if both the double bond substituents are held in a specific orientation. In the spectrum (Fig. 3–24)† of α-pinene (LXX), the resonance of

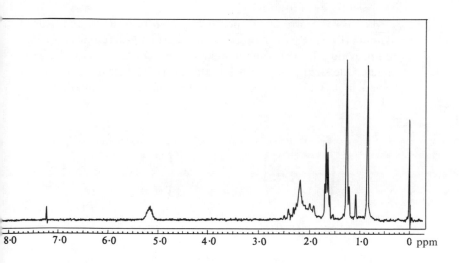

LXX

the vinyl methyl group (1·65) is split by one allylic interaction to *d* and two homoallylic interactions to the two *e* protons; all three splittings are equal (~2 Hz) and a 1:3:3:1 quartet therefore results. The assignment of resonances for protons *a*, *b*, and *d* should prove to be a simple exercise for the reader.

Fig. 3–24

† Varian catalogue, spectrum no. 272.

3–10. Tables of Coupling Constants

In addition to the tables of coupling constants already mentioned (Tables 3–8 to 3–12), Table 3–13 gives long-range, vicinal and geminal coupling constants in some common systems. Ranges of values for some useful ^{19}F—^1H and ^{31}P—^1H coupling constants are given in Tables 3–14 and 3–15. These Tables are appended to this chapter.

3–11. Shift Reagents

The amount of information that can be extracted from a conventional ^1H spectrum at 60 or 100 MHz is, in the case of complex molecules, often limited by the occurrence of overlapping resonances (particularly in the aliphatic (1–3 ppm) and aromatic (6·5–8 ppm) regions). This problem is exemplified by Fig. 3–6. Super-conducting magnets, which allow proton spectra to be obtained in the 200–300 MHz region, have been developed during the last few years. These magnets allow proton spectra to be obtained with better separation of previously overlapping signals (increased chemical shift), and with increased sensitivity, but the improvement is rarely dramatic and the instrumentation is very expensive to buy and operate. However, a cheap and efficient way to increase chemical shift differences has recently been found.

Many of the rare earth metals (e.g. europium (Eu) and praseodymium (Pr)) are paramagnetic and in certain β-diketone complexes can provide a local magnetic field suitable for use in NMR experiments. One such complex is Eu(DPM)$_3$ (LXXI), which can be dissolved in a CDCl$_3$ or CCl$_4$ solution of, say, an alcohol being

LXXI, Eu(DPM)$_3$ LXXII, M = Eu or Pr

$$CH_3(CH_2)_4CH_2OH + Eu(DPM)_3 \rightleftharpoons$$
$$CH_3(CH_2)_4CH_2O\cdots Eu(DPM)_3 \qquad (3\text{–}15)$$
$$|$$
$$H$$

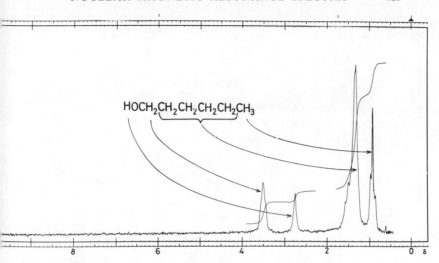

Fig. 3–25a

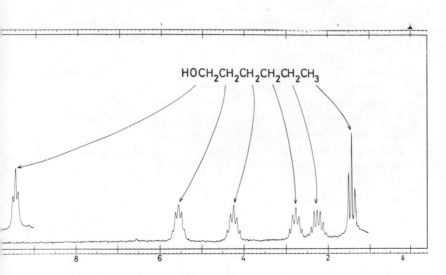

Fig. 3–25b
(superimposed trace offset 1 ppm)

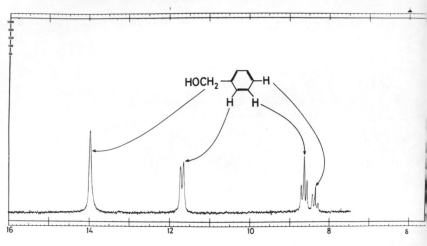

Fig. 3–26

examined in an NMR tube. The alcohol can then reversibly co-ordinate to the Eu(DPM)$_3$ at the OH-group (equation 3–15), and the local field due to the europium ion spreads out the spectrum. In the absence of angular variations, the shift falls off with the inverse cube of the distance from the metal, and the 100 MHz spectrum of n-hexanol (in CCl$_4$ solution, Fig. 3–25a) is spread out to that shown in Fig. 3–25b after the addition of Eu(DPM)$_3$ (0·29 moles of Eu(DPM)$_3$ per mole of substrate). The resonance due to the t-Bu groups of the complex does not interfere with spectral interpreta-tion since it lies near to that of TMS.

Similarly, the spectrum of benzyl alcohol appears as shown in Fig. 3–26 after the addition of Eu(DPM)$_3$ (0·39 moles). The signals due to each aromatic proton are clearly resolved, despite the fact that the normal NMR spectrum of benzyl alcohol shows only a single broad line in the aromatic region.

Eu(DPM)$_3$ normally causes downfield shifts in proton spectra, whereas Pr(DPM)$_3$ can be used to cause upfield shifts of similar magnitude. The ability of these reagents to shift spectra increases with increasing strength of the substrate as a Lewis base (Table 3–16 gives data for relatively unhindered common functional groups).

In polyfunctional molecules, each functional group competes for the shift reagent according to its relative ability to bind the reagent

Table 3–16

Eu(DPM)$_3$-Induced Shifts of Protons in some Common Environments

Functional group	Shift (Ppm/mole of Eu(DPM)$_3$ per mole of substrate)
RCH_2NH_2	~150
RCH_2OH	~100
RCH_2NH_2	30–40
RCH_2OH	20–25
RCH_2COR'	10–17
RCH_2CHO	19
RCH_2CHO	11
RCH_2OCH_2R	10
RCH_2CO_2Me	7
RCH_2CO_2Me	6·5
RCH_2CN	3–7

in monofunctional cases (Table 3–16 serves as a guide). The complexes Eu(FOD)$_3$ and Pr(FOD)$_3$ (see LXXII) are also commonly used to obtain more information from proton spectra.

3–12. Spin Decoupling

If two nuclei of spin $\frac{1}{2}$ interact in an AX system with a coupling constant J_{AX}, then four lines will be observed in the spectrum (Fig. 3–27a, $\delta_{AX} \gg J_{AX}$). However, if the A nucleus is strongly

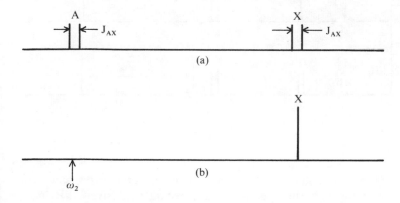

Fig. 3–27

irradiated at its resonance frequency by a radio frequency ω_2 while the signal due to the X nucleus is being observed by the normal observing radio frequency ω_1, then the X resonance is spin-decoupled to a singlet (Fig. 3–27b).

The spin decoupling can be regarded as arising because the radio frequency ω_2 causes such rapid transitions between the two possible spin states of the nucleus A that the X nucleus no longer 'sees' two distinct spin states of A. If A and X are different nuclei (e.g. ^{31}P and ^1H), then the experiment is known as heteronuclear decoupling and if A and X are nuclei of the same type (e.g. both ^1H) as homonuclear decoupling. These are special examples of *double resonance* experiments.

Homonuclear decoupling of proton spectra is used as a powerful method of structure elucidation in organic chemistry. The principle of the method is illustrated by the hypothetical 100 MHz NMR spectrum (Fig. 3–28A) of an 'unknown' compound $C_6H_8O_2$. Let us suppose that a frequency sweep double resonance experiment is

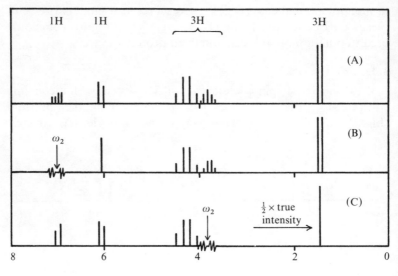

Fig. 3–28

now carried out, in which a decoupling frequency ω_2 is used to irradiate the signal near 7 ppm while the observing frequency ω_1 is swept through the spectrum (at constant field strength). In this experiment, a beat pattern can be observed near 7 ppm as ω_1 sweeps through ω_2 and, in addition, the doublet near 6 ppm is decoupled to a singlet and the quintet near 3·8 ppm is simplified to a 1:3:3:1 quartet (Fig. 3–28B). Making only the reasonable assumption that the one-proton signals near 7, 6, and 3·8 ppm are associated with protons attached to sp^2, sp^2, and sp^3 carbons, we can conclude that the system LXXIII is present.

$$
\begin{array}{ccc}
\sim 3\cdot 8 & \sim 7 & \sim 6 \\
\text{H} & \text{H} & \text{H} \\
-\text{C}-\text{C}=\text{C}-\text{X} \\
|
\end{array}
\qquad\qquad
\begin{array}{ccc}
\sim 3\cdot 8 & \sim 7 & \sim 6 \\
\text{H} & \text{H} & \text{H} \\
\text{CH}_3-\text{C}-\text{C}=\text{C}- \\
|
\end{array}
$$

LXXIII   LXXIV

If a second double resonance experiment is carried out, this time with the decoupling frequency ω_2 at 3·8 ppm, then the quartet near 7 ppm is simplified to a doublet (as expected), but in addition the doublet methyl resonance is simultaneously decoupled to a singlet (Fig. 3–28C). Therefore, we can extend our structural unit to LXXIV. The 6 protons which have been assigned are evidently not coupled to the 2 remaining protons, which hence must give an AB system near 4·2 ppm; J_{AB} is appreciably larger than any other coupling constant observed in this molecule and is consistent with geminal coupling, while the chemical shift and lack of further coupling are in accord with LXXV (where Y is not strongly electronegative and carries no directly bonded hydrogens). The gross chemical shifts, coupling constants and multiplicities are thus in accord with LXXVI (although it is emphasized that the spectrum is a hypothetical one).

Y—CH$_2$—O

LXXV

$$
\begin{array}{c}
\text{H} \qquad\qquad \text{H} \\
\diagdown \qquad\qquad \diagup \\
\text{C}=\text{C} \\
\diagup \qquad\qquad \diagdown \\
\text{CH}_3-\text{CH} \qquad\qquad \text{C}=\text{O} \\
\diagdown \qquad\qquad \diagup \\
\text{O}-\text{CH}_2
\end{array}
$$

LXXVI

3–13. ^{13}C NMR Spectra

The field strengths which allow proton spectra to be obtained at 100 MHz allow ^{13}C spectra to be recorded at 25·1 MHz. Due to the low sensitivity for recording ^{13}C spectra, pulsed spectra and Fourier transform techniques (section 3–2) are normally used. One advantage of ^{13}C magnetic resonance is the relatively large spread of chemical shifts (~ 200 ppm) which is observed (Table 3–17, appended to this chapter). Current practice is to measure chemical shifts relative to internal TMS or CS_2 ($\delta_{CS_2} = 192 - \delta_{TMS}$ ppm). It is noteworthy that, in analogy to proton spectra, ^{13}C resonances of hydrocarbons lie close to TMS, those of sp^3 carbons in ^{13}C—O and ^{13}C—N groups lie at lower field, and the sp^2 carbons of double bonds and aromatic rings are displaced even further (>90 ppm) from TMS; the 'left hand end' of the spectra (as conventionally displayed) contain the regions in which carbonyl carbons resonate.

Resonances due to ^{13}C nuclei directly attached to one, two, or three equivalent hydrogens would normally exhibit large doublet, triplet, or quartet splittings, since directly bonded ^{13}C—^1H coupling constants are large (e.g., CH_3CH_3, 125 Hz; CH_3OH, 141 Hz; $CH_2{=}CH_2$, 150 Hz; $H_2C{=}O$, 172 Hz; $HC{\equiv}CH$, 249 Hz; $HC{\equiv}N$, 269 Hz). Longer range couplings are mainly small (0–10 Hz), but nevertheless it is apparent that a ^{13}C spectrum could be very complicated. However, the multiplicities due to all ^{13}C—^1H couplings can be removed by irradiating the whole of the proton spectrum while the ^{13}C spectrum is recorded. This technique is called proton noise decoupling and provides an example of heteronuclear decoupling (section 3–12). With proton noise decoupling, ^{13}C spectra (if recorded under the most favourable conditions) exhibit one signal for each unique carbon atom present in the sample. This point is

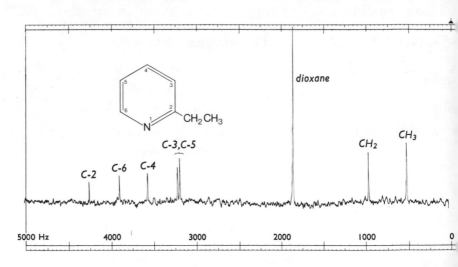

Fig. 3–29

illustrated by the 25·1 MHz ^{13}C Fourier transform NMR spectrum (Fig. 3–29) of 2–ethylpyridine in dioxane (30 per cent vol./vol.).

Pulsed Fourier transform ^{13}C magnetic resonance is a relatively new technique, and this fact, in conjunction with the expense of the instrumentation, currently precludes its use as a routine method available to all. Yet the power of the method is so great that it seems likely that within a decade it will be as widely used as proton magnetic resonance.

Table 3–1

Chemical Shifts of Methyl, Methylene and Methine Protons†

Methyl protons			Methylene protons			Methine protons		
Proton	δ	τ	Proton	δ	τ	Proton	δ	τ
CH_3—C	0·9	9·1	—C—CH_2—C	1·4	8·6	C—CH—C	1·5	8·5
CH_3—C—C=C	1·1	8·9	—C—CH_2—C—C=C	1·7	8·3			
CH_3—C—O	1·3	8·7	—C—CH_2—C—O	1·9	8·1	—C—CH—C—O	2·0	8·0
CH_3—C=C	1·6	8·4	—C—CH_2—C=C	2·3	7·7			
CH_3—Ar	2·3	7·7	—C—CH_2—Ar	2·7	7·3	—CH—Ar	3·0	7·0
CH_3—CO—R	2·2	7·8	—C—CH_2—CO—R	2·4	7·6	—C—CH—CO—R	2·7	7·3
CH_3CO—Ar	2·6	7·4				—C—CH—CO-Ar	3·3	6·7
CH_3—CO—O—R	2·0	8·0	—C—CH_2—CO—O—R	2·2	7·8			
CH_3—CO—O—Ar	2·4	7·6						
CH_3—CO—N—R	2·0	8·0	—C—CH_2—CO—N—R	2·2	7·8			
CH_3—O—R	3·3	6·7	—C—CH_2—O—R	3·4	6·6	—C—CH—O—R	3·7	6·3
			—C—CH_2—O—H	3·6	6·4	—C—CH—OH	3·9	6·1
CH_3—O—C=C	3·8	6·2						
CH_3—O—Ar	3·8	6·2	—C—CH_2—O—Ar	4·3	5·7			
CH_3—O—CO—R	3·7	6·3	—C—CH_2—O—CO—R	4·1	5·9	—C—CH—O—CO—R	4·8	5·2
CH_3—N	2·3	7·7	—C—CH_2—N	2·5	7·5	—C—CH—N	2·8	7·2
CH_3—$\overset{+}{N}$	3·3	6·7						
CH_3—N—Ar	3·0	7·0						
CH_3—S	2·1	7·9	—C—CH_2—S	2·4	7·6	—C—CH—S	3·2	6·8
			—C—CH_2—NO_2	4·4	5·6	—C—CH—NO_2	4·7	5·3

Group			Group			Group		
CH_3—C—NO_2	1·6	8·4	—C—CH_2—C—NO_2	2·1	7·9	C—CH—Br	4·3	5·7
CH_3—C=C—CO	2·0	8·0	—C—CH_2—C=C—CO	2·4	7·6	C—CH—I	4·3	5·7
C=C(CH_3)—CO	1·8	8·2	C=C(CH_2)—CO	2·4	7·6	C—CH—C≡N	2·7	7·3
			C—CH_2—Cl	3·6	6·4	C—CH—N—CO—R	4·1	5·9
			C—CH_2—Br	3·5	6·5			
			C—CH_2—I	3·2	6·8			
			C—CH_2—C≡N	2·3	7·7			
CH_3—N—CO—R	2·9	7·1		5·9	4·1			

† Reproduced with permission from K. Nakanishi, *Infrared Absorption Spectroscopy*, Appendix 1, p. 223 (Holden-Day, San Francisco and Nankado Company Ltd., Tokyo, 1962), but with additional data added. Usually values will be within ±0·2 ppm of the values quoted unless inductive or steric effects associated with additional groups are operative.

Table 3–2

Estimation of Chemical Shift (δ) for Protons of —CH$_2$— and —$\overset{|}{C}$H— Groups

$$\delta_{CH_2} = 0{\cdot}23 + C_1 + C_2 \qquad \delta_{CH} = 0{\cdot}23 + C_1 + C_2 + C_3$$

X	C	X	C	X	C
—CH$_3$	0·5	—SR	1·6	—OR	2·4
—CF$_3$	1·1	—C≡C—Ar	1·7	—Cl	2·5
>C=C<	1·3	—CN	1·7	—OH	2·6
—C≡C—R	1·4	—CO—R	1·7	—N=C=S	2·9
—COOR	1·5	—I	1·8	—OCOR	3·1
—NR$_2$	1·6	—Ph	1·8	—OPh	3·2
—CONR$_2$	1·6	—Br	2·3		

Table 3–3

δ-Values of Acidic Protons†

	δ(ppm)		δ(ppm)
ROH	0·5–4·5	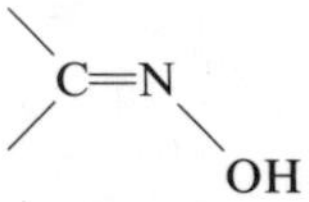 >C=N—OH	9–12
RNH$_2$, RNHR	1–5	RCOOH	9–13
RSH	1–2	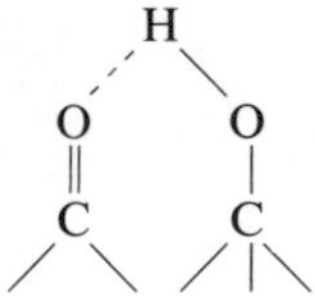	
ArOH	4·5–6·5	(H bonded O=C···O—C)	7–13
ArNH$_2$, ArNHR	3–6	RCONH$_2$ } RCONHR }	5–12
ArSH	3–4		

† Shifts vary widely with temperature, concentration, pH and solvent. Signals often broadened. Low field shifts caused by hydrogen bonding.

Table 3–4

Chemical Shifts of Ortho, Meta, and Para Protons in Monosubstituted Benzenes (in ppm from benzene, $\delta = 7.27$ ppm)†

Substituent	Δ_{ortho}	Δ_{meta}	Δ_{para}
NO_2	0·94	0·18	0·39
CHO	0·58	0·20	0·26
COOH	0·80	0·16	0·25
$COOCH_3$	0·71	0·08	0·20
COCl	0·82	0·21	0·35
CCl_3	0·8	0·2	0·2
$COCH_3$	0·62	0·10	0·25
CN	0·26	0·18	0·30
$CONH_2$	0·65	0·20	0·22
$\overset{+}{N}H_3$	0·4	0·2	0·2
$CH_2X‡$	0·0–0·1	0·0–0·1	0·0–0·1
CH_3	−0·16	−0·09	−0·17
CH_2CH_3	−0·15	−0·06	−0·18
$CH(CH_3)_2$	−0·14	−0·09	−0·18
$C(CH_3)_3$	−0·09	0·05	−0·23
F	−0·30	−0·02	−0·23
Cl	0·01	−0·06	−0·08
Br	0·19	−0·12	−0·05
I	0·39	−0·25	−0·02
NH_2	−0·76	−0·25	−0·63
OCH_3	−0·46	−0·10	−0·41
OH	−0·49	−0·13	−0·2
OCOR	−0·2	0·1	−0·2
$NHCH_3$	−0·8	−0·3	−0·6
$N(CH_3)_2$	−0·60	−0·10	−0·62

† A positive Δ value indicates a shift to low field relative to benzene.
‡ X = Cl, alkyl, OH, or NH_2.

Table 3–5A

Approximate Chemical Shifts of Protons attached to Unsaturated Linkages

Proton	δ	τ	Proton	δ	τ
R—CHO	9·4–10·0	0·0–0·6	—C=CH—	4·5–6·0	4·0–5·5
Ar—CHO	9·7–10·5	−0·5–0·3	—C=CH—CO	5·8–6·7	3·3–4·2
H—CO—O	8·0–8·2	1·8–2·0	—CH=C—CO	6·5–8·0	2·0–3·5
H—CO—N	8·0–8·2	1·8–2·0	—CH=C—O	4·0–5·0	5·0–6·0
—C≡C—H	1·8–3·1	6·9–8·2	—C=CH—O	6·0–8·1	1·9–4·0
Aromatic			—CH=C—N	3·7–5·0	5·0–6·3
protons	6·0–9·0	1·0–4·0	—C=CH—N	5·7–8·0	2·0–4·3

Table 3–5B

Chemical Shifts (δ ppm) of Protons in some Unsaturated Cyclic Systems

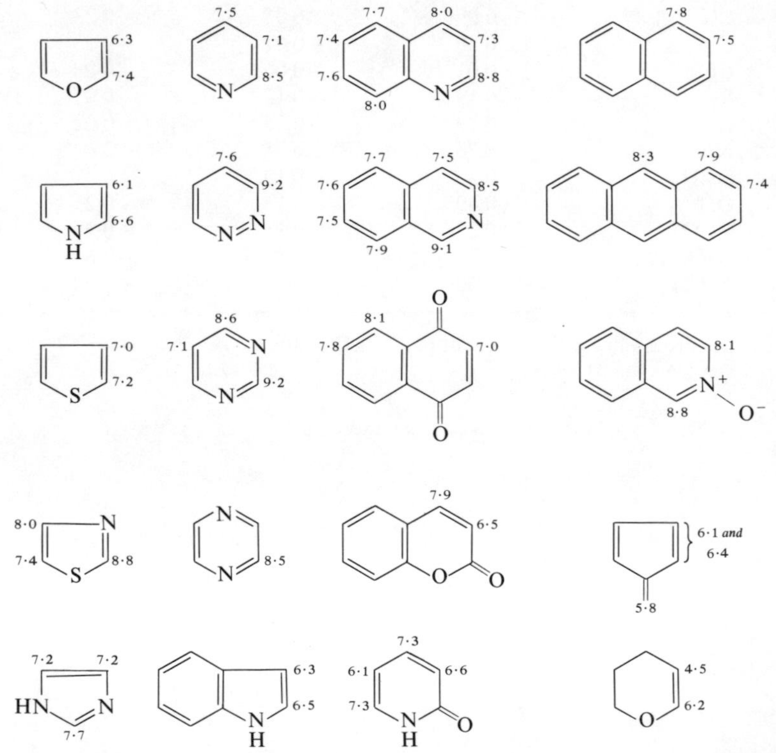

Table 3–6

Estimation of Chemical Shift of a Proton Attached to a Double Bond†‡

$$\delta_{C=C \atop H} = 5.25 + Z_{gem} + Z_{cis} + Z_{trans}$$

R	Z_i for R (ppm)		
	Z_{gem}	Z_{cis}	Z_{trans}
→H	0	0	0
→Alkyl	0·45	−0·22	−0·28
→Alkyl—Ring	0·69	−0·25	−0·28
→CH$_2$O—	0·64	−0·01	−0·02
→CH$_2$S—	0·71	−0·13	−0·22
→CH$_2$X(X: F, Cl, Br)	0·70	0·11	−0·04
→CH$_2$N⟨	0·58	−0·10	−0·08
C=C (isolated)	1·00	−0·09	−0·23
C=C (conjugated)	1·24	0·02	−0·05
→C≡N	0·27	0·75	0·55
→C≡C—	0·47	0·38	0·12
C=O (isolated)	1·10	1·12	0·87
C=O (conjugated)	1·06	0·91	0·74
→COOH (isolated)	0·97	1·41	0·71
→COOH (conjugated)	0·80	0·98	0·32
→COOR (isolated)	0·80	1·18	0·55
→COOR (conjugated)	0·78	1·01	0·46
H—→C=O	1·02	0·95	1·17
N—→C=O	1·37	0·98	0·46

Table 3–6 *continued*

R	Z_i for R (ppm)		
	Z_{gem}	Z_{cis}	Z_{trans}
$\overset{\displaystyle Cl}{\underset{\displaystyle \mid}{\rightarrow}}C{=}O$	1·11	1·46	1·01
→OR (R: aliphatic)	1·22	−1·07	−1·21
→OR (R: conjugated)	1·21	−0·60	−1·00
→OCOR	2·11	−0·35	−0·64
→CH$_2$—$\overset{\mid}{C}{=}$O; →CH$_2$—C≡N	0·69	−0·08	−0·06
→CH$_2$—Aromatic—Ring	1·05	−0·29	−0·32
→Cl	1·08	0·18	0·13
→Br	1·07	0·45	0·55
→I	1·14	0·81	0·88
→$\overset{\mid}{N}$—R (R: aliphatic)	0·80	−1·26	−1·21
→$\overset{\mid}{N}$—R (R: conjugated)	1·17	−0·53	−0·99
→$\overset{\mid}{N}$—$\overset{\mid}{C}{=}$O	2·08	−0·57	−0·72
→Aromatic	1·38	0·36	−0·07
→SR	1·11	−0·29	−0·13
→SO$_2$	1·55	1·16	0·93

† Taken from U. E. Matter, C. Pascual, E. Pretsch, A. Pross, W. Simon and S. Sternhell, *Tetrahedron*, **25**, 691 (1969). Positive Z values indicate a downfield shift and an arrow (→) indicates the point of attachment of R to the double bond.

‡ The increments 'R conjugated' are used instead of 'R isolated' when either the substituent or the double bond is conjugated with further substituents. The increments 'Alkyl—Ring' are used when the substituent together with the double bond are a part of a 5- or 6-membered cyclic structure.

Table 3–7

Chemical Shifts (δ ppm) of Protons in some Saturated Heterocyclic Ring Systems

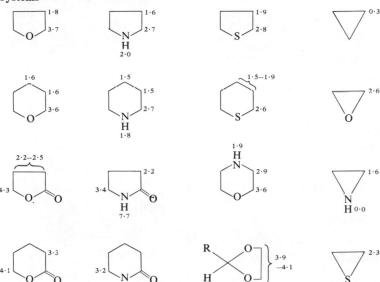

Table 3–10

Vicinal Proton Coupling Constants (Hz) Observed in some Unsaturated Heterocyclic Systems

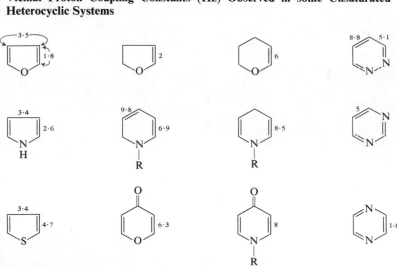

Table 3–13

Spin–Spin Coupling Constants in some Common Systems†

System	$J_{H,H'}$ Full Range	$J_{H,H'}$ Typical	System	$J_{H,H'}$ Full Range	$J_{H,H'}$ Typical
$>C<^H_{H'}$	0–25	10–15	$>C=C<^{H\ \ H'}$ (cis)	0–12	7–10
$>CH-CH'<$	0–8	~7	$>C=C<^{H'}$ (trans, H below)	12–18	14–16
$CH_3-CH_2'-$ Free rotation	6–8	~7	aromatic $J_{H,H'}$	6–10	8
$CH_3\!\!>CH'-$ Free rotation	5–7	~6	aromatic $J_{H,H''}$	0–3	2
$>CH-\underset{\mid}{\overset{\mid}{C}}-CH'<$	0–1	0	aromatic $J_{H,H'''}$	0–1	1
$>C=CH-CH'<$	4–10	5–7	$=C<^H_{H'}$	0–3.5	2
$>C=CH-CH'=C<$	6–13	10–13	$-CH=\underset{\mid}{C}-CH'<$	0–3	0.5–2
$>CH-CH'O$	0–3	2	$>CH-\underset{\mid}{C}=\underset{\mid}{C}-CH'<$	0–2	1
$C=CH-CH'O$	5–8	7	$>CH-C≡CH'$	2–3	2.5

† For more detailed values, see text.

Table 3–14

^{19}F—^{1}H Coupling Constants

Type of Coupling	Structure	Range (Hz)
$^2J_{HF}$		45–52
$^2J_{HF}$		60–65
$^2J_{HF}$		72–90
$^3J_{HF}$	$CH_3{-}CF$	20–24
$^3J_{HF}$	$HC{-}CF$	0–45 ($J_{gauche} = 0\text{–}12$) ($J_{trans} = 10\text{–}45$)
$^3J_{HF}$		3–20
$^3J_{HF}$		12–53
J_{ortho} J_{meta} J_{para}		6–11 3–9 0–4
J_{ortho} J_{meta} J_{para}		2·5 1·5 0

Table 3–14 *continued*

Type of Coupling	Structure	Range (Hz)
$^4J_{HF}$	HC—C—CF	0–9†
$^4J_{HF}$	$\begin{matrix} & \diagdown \\ & C=C \\ F & \diagup \diagdown \\ & CH_3 \end{matrix}$	2–4
$^4J_{HF}$	$\begin{matrix} & CF \\ & \diagdown & \diagup \\ & C=C \\ H & \diagup & \diagdown \end{matrix}$	0–6

† Larger values ($^4J_{HF} \geqslant 3\cdot5$ Hz) within the quoted range observed for

$\begin{matrix} H & & C & & F \\ \diagdown & \diagup & \diagdown & \diagup \\ & C & & C \end{matrix}$

configuration of nuclei.

Table 3–15

^{31}P–^{1}H **Coupling Constants (Hz)**†

Type of Coupling	Class of Compound		
	Phosphines	Phosphonium salts	Phosphine oxides
$^1J_{PH}$	(150) 185–220 (250)	400–900	200–750
$^2J_{PH}$	(−5) 0–15 (27) 46‡	(0) 10–18 30‡	5–25 40‡
$^3J_{PCCH}$	(10) 13–17 (20)	(0) 10–20 (57)	14–30
$^3J_{PC=CH}$	*trans* (5) 12–41 *cis*§ 6–20	*trans* 28–50 (80) *cis*§ (2) 10–20 (35)	
		Phosphites	Phosphates
$^3J_{POCH}$		(0) 5–14 (20)	(0) 5–20 (30)
	All compounds		
$^4J_{PH}$	0–3 (5)*		

† The coupling constants are often strongly dependent upon the groups attached to phosphorus, and therefore values outside the quoted ranges may occasionally be observed; values in parentheses are 'extreme' values so far reported.

‡ Values observed in P—$\overset{\overset{\displaystyle C}{\|}}{C}$—H systems.

§ *trans*-coupling is usually about twice that of *cis*-coupling.

* In the system P—C=C=CH.

Table 3-17
Chemical Shifts of some ¹³C Resonances (relative to internal TMS; δ_{TMS} equals zero ppm)

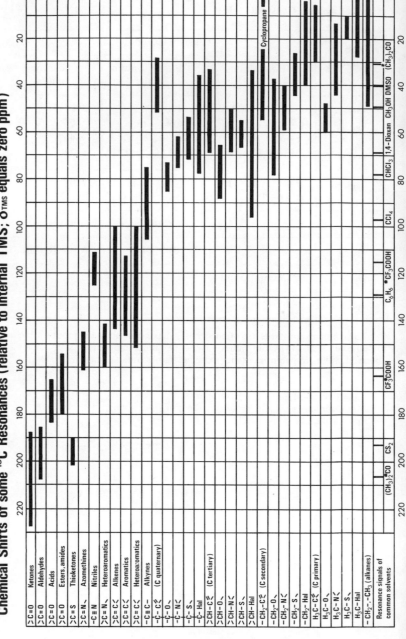

Table 3–18

Solvent Positions of Residual Protons in Incompletely Deuterated Solvents

Solvent	Group	δ (ppm)
d_4-Acetic acid	Methyl	2·05
	Hydroxyl	11·5†
d_6-Acetone	Methyl	2·05
d_3-Acetonitrile	Methyl	1·95
d_6-Benzene	Methine	7·3
d_1-t-Butanol (OD)	Methyl	1·28
d_1-Chloroform	Methine	7·25
d_{12}-Cyclohexane	Methylene	1·40
Deuterium oxide	Hydroxyl	4·7†
d_7-Dimethylformamide	Methyl	2·75
	Methyl	2·95
	Formyl	8·05
d_6-Dimethylsulphoxide	Methyl	2·5
	Absorbed water	3·3†
d_8-Dioxan	Methylene	3·55
d_{18}-Hexamethylphosphoramide	Methyl	2·60, d, J = 9 Hz
d_4-Methanol	Methyl	3·35
	Hydroxyl	4·8†
d_2-Methylene chloride	Methylene	5·35
d_5-Pyridine	C—2 Methine	8·5
	C—3 Methine	7·0
	C—4 Methine	7·35
d_8-Toluene	Methyl	2·3
	Methine	7·2
d_1-Trifluoroacetic acid	Hydroxyl	11·3†

† These values may vary greatly, depending upon the solute and its concentration.

Bibliography

J. A. Pople, W. G. Schneider and H. J. Berstein, *High Resolution Nuclear Magnetic Resonance*, McGraw-Hill, New York, 1959.

J. D. Roberts, *Nuclear Magnetic Resonance*, McGraw-Hill, New York, 1959.

L. M. Jackman, *Applications of NMR Spectroscopy in Organic Chemistry*, Pergamon, London, 1959.

The NMR–EPR staff of Varian Associates, *NMR and EPR Spectroscopy*, Pergamon, New York, 1960.

J. D. Roberts, *An Introduction to the Analysis of Spin–Spin Splitting in Nuclear Magnetic Resonance*, Benjamin, New York, 1961.

K. B. Wiberg and B. J. Nist, *The Interpretation of NMR Spectra*, Benjamin, New York, 1962.

J. W. Emsley, J. Feeney and L. H. Sutcliffe, *High Resolution Nuclear Magnetic Spectroscopy*, Vols. 1 and 2, Pergamon, London, 1965 and 1966.

N. S. Bhacca and D. H. Williams, *Applications of NMR Spectroscopy in Organic Chemistry*, Holden-Day, San Francisco, 1964.

J. W. Emsley, J. Feeney and L. H. Sutcliffe (eds.), *Progress in NMR Spectroscopy*, Vols. 1–5, Pergamon, London, 1966–1969.

E. F. Mooney (ed.), *Annual Review of NMR Spectroscopy*, Vols. 1–4, Academic Press, London, 1968–1971.

L. M. Jackman and S. Sternhell, *Applications of Nuclear Magnetic Resonance Spectroscopy in Organic Chemistry*, Pergamon, London, 1969.

T. C. Farrar and E. D. Becker, *Pulse and Fourier Transform NMR*, Academic Press, New York, 1971.

J. B. Stothers, *Carbon-13 N M R Spectroscopy*, Academic Press, New York, 1972.

G. C. Levy and G. L. Nelson, *Carbon-13 Nuclear Magnetic Resonance for Organic Chemists*, Wiley-Interscience, 1972.

Catalogues and Literature Citations

NMR Spectra Catalogue, Volumes 1 and 2, Varian Associates, Palo Alto, California.

Nuclear Magnetic Resonance Spectral Data, Manufacturing Chemists Association Research Project, Department of Chemistry, Texas A & M University, College Station, Texas.

M. G. Howell, A. S. Kende and J. S. Webb, *Formula Index to NMR Literature Data*, Vol. 1: References prior to 1961, Plenum Press, New York, 1965.

NMR, NQR, EPR Current Literature Service, Butterworths and Verlag Chemie; covers the literature from the beginning of 1963.

L. F. Johnson and W. C. Jankowski, *Carbon-13 NMR Spectra*, Wiley, New York, 1972.

4. Mass Spectra

4–1. Introduction

The extensive application of mass spectrometry in organic chemistry only began around 1960. There has been a great increase in the use of this physical method since that time. There are two main reasons for the burst of activity in this field. First, it is now generally appreciated by organic chemists that there are a number of instruments available which are able to volatilize the vast majority of organic compounds in which they are interested (most usually of molecular weight less than 1000, but not necessarily so), to ionize the vapour, and hence to record the molecular weight by measurement of the mass to charge ratio (m/e value). Second, it is also appreciated that the molecular ion, formed as described above, breaks down into charged fragments whose structures (as deduced from their m/e values) can be related to the structure of the intact molecule. It is the purpose of this chapter to illustrate briefly how the mass spectra of organic compounds can be obtained and presented and, most important, how they can be interpreted.

4–2. The Mass Spectrometer

A. The Ion Chamber and Ionization

In discussing the basic principles of the mass spectrometer, let us first assume that we have surmounted the problem of introducing a few micrograms of vapour of the sample into the high vacuum system (at *ca.* 10^{-6} mm. Hg) of the spectrometer. The vapour is allowed to pass through a slit A into the ion chamber (Fig. 4–1), where it is bombarded by a beam of electrons accelerated from a filament, usually to an energy of about 70 eV.† One of the processes induced by the electron bombardment is ionization of the molecules of vapour by removal of one electron; a positively charged molecular ion (a) is formed. Since organic molecules are almost without exception even-electron species, the process of removing one electron leads to a radical-ion containing one unpaired electron.

$$[M] \xrightarrow{-e} [M]^{\ddot{+}} \qquad\qquad [M] \xrightarrow{+e} [M]^{\ddot{-}}$$
$$a \qquad\qquad\qquad\qquad b$$

The alternative process, involving capture of an electron by a molecule of vapour to afford a negative radical-ion (b) is less probable by a factor of about 10^{-2}. In this chapter we will be concerned only with positive ion mass spectrometry.

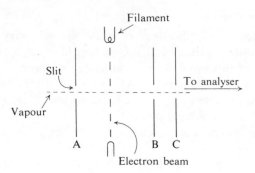

Fig. 4–1

The energy of an electron beam which is required to remove an electron from organic molecules is usually of the order of 10 eV. Therefore virtually no ions are produced if the energy of the beam is much lower than 10 eV. However, if the energy of the bombarding electrons is much greater than 10 eV (e.g., 70 eV), then the additional energy of the electrons may be dissipated in breaking bonds in the molecular ion. In this way, fragment ions are also produced.

The various positive ions generated by electron impact are then accelerated through a second slit by a repeller potential applied between A and B. Finally, a large accelerator potential (of the order

† 1 eV $\simeq$ 23 kcal./mole $\simeq$ 96 kJ/mole.

of 8 kV) is applied between B and C, and the positive ions travel with a high velocity into the analyser portion of the mass spectrometer, where they will be separated according to their m/e ratios.

B. Magnetic and Electrostatic Analysers. High Resolution

If we wish merely to separate all ions in the spectrometer which differ by at least unit mass (e.g., to resolve m/e 110 from m/e 111, where these values represent singly charged fragments whose atomic constituents add up to masses of 110 and 111 respectively), it is sufficient to deflect the ions only in a strong magnetic field. Ions of larger mass are deflected less than ions of smaller mass according to equation 4–1, where H is the strength of the magnetic field, r is the radius of the circular path in which the ion is travelling, and V is the accelerating potential.

$$\frac{m}{e} = \frac{H^2 r^2}{2V} \qquad (4\text{-}1)$$

The radial paths followed by ions in a magnetic field are illustrated in Fig. 4–2. It will be obvious that by scanning the magnetic field, equation 4–1 can be satisfied for ions of all m/e ratios for fixed values of r and V. Alternatively, the mass spectrum may be scanned electrically by varying V while the magnetic field is held constant. Whichever device is employed, ions of all m/e values can be successively allowed to pass through the collector slit D (Fig. 4–2) and the mass spectrum recorded.

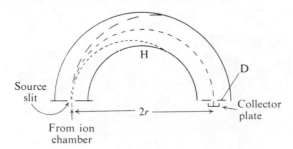

Fig. 4–2

Frequently, we may wish to differentiate between ions which have the same nominal integral mass, but possess different exact masses. This should be possible since, generally speaking, the isotopes of which elements consist do not have exact integral masses. Based on the convention that the atomic weight of ^{12}C is 12 exactly, the masses of the most abundant isotopes of hydrogen, nitrogen and oxygen are given in Table 4–1. Hence it will be evident that although CO, H_2CN, $CH_2{=}CH_2$ and N_2 all have the same integral mass (28), the exact masses of the four species are different as indicated in the table.

Table 4–1

Exact Masses of Some Common Isotopes and Simple Molecular Species

Species	^1H	^{16}O	^{14}N	CO	H_2CN	$CH_2{=}CH_2$	N_2
Mass	1·00782	15·9949	14·0031	27·9949	28·0187	28·0313	28·0061

By using a high resolution mass spectrometer, it is possible to separate the positive ions corresponding to CO, H_2CN, $CH_2{=}CH_2$ and N_2. The tremendous advantages of high resolution mass spectrometry will be mentioned again in section 4–11, but here we will be concerned with the manner in which such high resolving power is achieved. Normally, the ions leaving the ion source will have a spread of energies, because of variable thermal energies possessed prior to acceleration, and due to penetration of the accelerating field into the ion source. If mono-energetic ions can be selected for separation in the magnetic analyser, much more accurate focusing at the collector plates can be achieved. In practice, mono-energetic ions are selected for magnetic analysis by means of an electrostatic analyser. The radial electrostatic field of the electro-

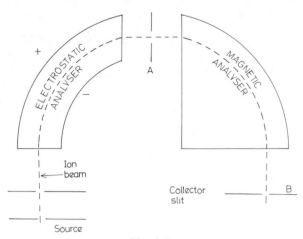

Fig. 4–3

static analyser effects velocity focusing of the various ions at a slit A (Fig. 4–3). For any given m/e value, the mono-energetic ions can then be very accurately focused at B (the collector slit), and ions whose masses only differ by as little as a few parts per million can in fact be separated, if necessary.

Typically, it takes an ion 1–5 μsec. after its formation to leave an electron impact source, and about 1 μsec. to traverse the field-free region in front of the magnetic analyser. In a double focusing instrument (Fig. 4–3), this field-free region is reached after about 10 μsec. for an ion of m/e 100 at an accelerating voltage of 8 kV; the total time taken to travel from the source to collector is about 20 μsec.

C. Inlet Systems

We will now consider briefly some of the systems which enable us to introduce the vapour of an organic compound into the ion chamber of the spectrometer. If the substance we wish to study is a gas or a volatile liquid, the inlet system illustrated schematically in Fig. 4–4 may be used.

If the spectrometer is not being used to examine a sample, valve C may be kept closed; and when valves A and B are open, all parts of the inlet system above C are under vacuum. A volatile liquid sample in the ampoule may be retained in the ampoule by cooling with liquid nitrogen while air is removed with A and C open, and B closed. Once the whole inlet system is under vacuum, A is closed, B opened and the liquid nitrogen which is cooling the ampoule is removed; the sample then volatilizes into the reservoir from which it can diffuse at a suitable rate through a sinter to the ion chamber.

A large number of organic compounds are relatively volatile and thermally stable solids with molecular weights in the approximate range of 100 to 300. Such samples may be introduced into the ion chamber by means of a system analogous to that illustrated in Fig. 4–4, but having the whole (except the sample ampoule) encased in an oven whose temperature can be varied in the range from room temperature to 300°C; the sample ampoule can, of course, be heated separately. In this way, adequate vapour pressures can be obtained from relatively volatile solids. Sample size is of the order of 1 mg. for such inlet systems.

Finally, we often wish to deal with samples which have insufficient thermal stability to be heated to 200–300° or have very low vapour pressures even at these temperatures. Many substances

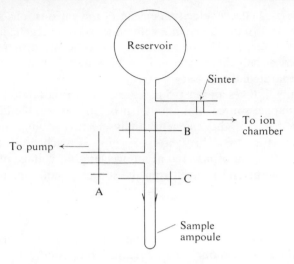

Fig. 4–4

in the molecular weight range 300 to 1200 fall into the latter class. For these cases, advantage is taken of the extremely low pressures which can be achieved in the ion chamber. The sample is introduced directly into the ion chamber on the end of a probe. The end of the probe may be heated because of its proximity to the electron beam or, in addition, by means of a heater wire in the probe. A vapour pressure of about 10^{-6} mm. Hg permits a spectrum to be obtained using this direct insertion technique on a sample of a few μg. In favourable cases useful information can be obtained from samples in the ng. to pg. range (10^{-9} to 10^{-12} g.). The high sensitivity of the method is one of its great advantages.

4–3. Representation of Mass Spectra

Mass spectra may be scanned from low to high mass or from high to low mass. The recorders employed in mass spectroscopy differ from those usually employed in optical or NMR spectroscopy because tremendous variations in the abundances of ions which constitute a mass spectrum are observed, and because ions of very low relative abundance may assist a structure elucidation. A common procedure which is used to surmount this problem employs a series of mirror galvanometers with increasing sensitivities, e.g., six galvanometers may be used with sensitivity ratios of 1:3:10:30:100:300. The galvanometers are deflected when ions impinge on the collector plate and a beam of ultraviolet light can therefore be made to trace out a series of peaks on a chart of ultraviolet sensitive paper driven by a conventional motor drive. The partial mass spectrum (above m/e 110) of diethyl 2-acetylglutarate (**1**) as it appears when using such a recorder (with three galvanometers) is illustrated in Fig. 4–5. While the device permits us to

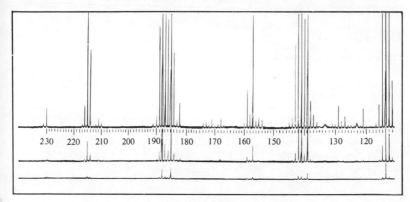

Fig. 4–5

detect occasional very small but significant peaks, it is obviously highly inconvenient if we wish to comprehend the spectrum at a glance. A clear representation of the main features may be obtained by plotting m/e values against relative abundance, arbitrarily assigning the most abundant ion (base peak) in the spectrum as

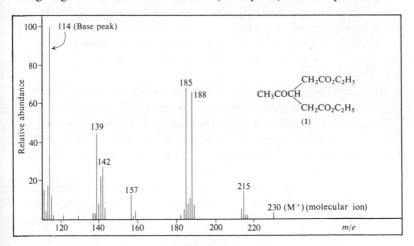

Fig. 4–6

100 per cent (Fig. 4–6). Alternatively, the same data may be represented in tabular form (Table 4–2). The clear visual presentation (Fig. 4–6) is analogous to conventional UV, IR and NMR presentations, and is employed in this book since tables cannot be comprehended at a glance. Other possible recording devices (e.g., photographic plates, attenuated pen-and-ink recorders) are beyond the scope of this book.

Table 4–2

Partial Mass Spectrum (above m/e 110) of Diethyl 2-Acetylglutarate (1)

m/e value		111	112	113	114	115	116	121	129	137
Relative abundance		15	3	18	100	12	2	2	2	3

m/e	138	139	140	141	142	143	157	158	159	182	184
R.A.	3	45	8	22	27	6	12	1	3	2	5

m/e	185	186	187	188	189	214	215	216	217	230
R.A.	67	8	11	66	7	5	14	2	2	3

4–4. Isotope Abundances

All singly charged ions in the mass spectrum which contain carbon also give rise to a peak at one mass unit higher. This happens because of the natural abundance of ^{13}C (1·1 per cent). For an ion containing n carbon atoms, the abundance of the isotope peak is $n \times 1·1$ per cent of the ^{12}C-containing peak. Thus $C_5H_{12}{}^+$, $C_{40}H_{70}{}^+$ and $C_{100}H_{170}{}^+$ would give isotope peaks at one mass unit higher of approximate abundances 5·5, 44 and 110 per cent of the abundance of the ions containing ^{12}C only. Obviously, the probability of finding two ^{13}C atoms in an ion is very low and M + 2 peaks are accordingly of very low abundance.

Although iodine and fluorine are monoisotopic, chlorine consists of ^{35}Cl and ^{37}Cl in the ratio of approximately 3:1 and bromine of ^{79}Br and ^{81}Br in the ratio of approximately 1:1. Molecular ions (or fragment ions) containing various numbers of chlorine and/or bromine atoms therefore give rise to the patterns shown in Fig. 4–7 (all peaks spaced 2 mass units apart).

Obviously, the isotope patterns to be expected from any combination of elements can readily be calculated, and provide a useful test of ion composition in those cases where polyisotopic elements are involved. Most of the remaining elements in Table 4–3 are essentially monoisotopic, with the exception of sulphur and silicon.

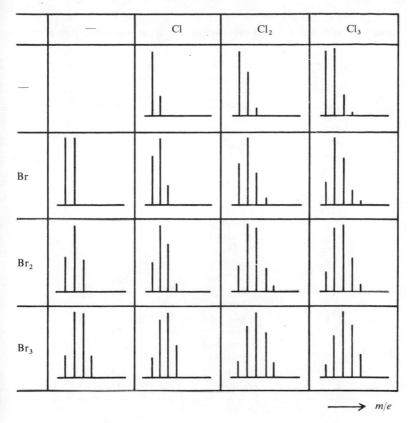

	—	Cl	Cl₂	Cl₃

Fig. 4–7

4–5. Factors Controlling Fragmentation Modes: Metastable Ions

The minimum energy required if a bombarding electron is to cause ionization of a molecule is the ionization potential (I.P.) of that molecule. This is the energy needed to remove an electron from the highest occupied molecular orbital of the molecule. In the case of a beam of 70 eV electrons, some of the electrons interact weakly with the molecule to be ionized and supply only the minimum energy (*ca.* 10 eV) needed for ionization. However, others interact more strongly, remove electrons from lower-lying molecular orbitals and produce initially a variety of electronically excited ions. In most cases the excess energy initially present as electronic energy

Table 4–3

Atomic Weights and Approximate Natural Abundance of Some Isotopes

Isotope	Atomic Weight ($^{12}C = 12 \cdot 000000$)	Natural Abundance (per cent)
^1H	1·007825	99·985
^2H	2·014102	0·015
^{12}C	12·000000	98·9
^{13}C	13·003354	1·1
^{14}N	14·003074	99·64
^{15}N	15·000108	0·36
^{16}O	15·994915	99·8
^{17}O	16·999133	0·04
^{18}O	17·999160	0·2
^{19}F	18·998405	100
^{28}Si	27·976927	92·2
^{29}Si	28·976491	4·7
^{30}Si	29·973761	3·1
^{31}P	30·973763	100
^{32}S	31·972074	95·0
^{33}S	32·971461	0·76
^{34}S	33·967865	4·2
^{35}Cl	34·968855	75·8
^{37}Cl	36·965896	24·2
^{79}Br	78·918348	50·5
^{81}Br	80·916344	49·5
^{127}I	126·904352	100

is probably converted into vibrational energy of the electronic ground states before fragmentation. Thus, as a first approximation, we may consider that in the ion source, prior to fragmentation, the ions have a distribution of energies, extending from almost zero internal energy to a maximum of $E_{el} + E_{th} - $ I.P., where E_{el} is the electron beam energy and E_{th} is the maximum thermal energy of the molecules prior to ionization. Since the various energy transfers occur with different probabilities, a plot of the probability $[P(E)]$ of an ion possessing an internal energy E against that energy takes the general form shown in Fig. 4–8a.

Since the source operating pressure is *ca.* 10^{-6} mm. Hg, there are very few molecular collisions and the energy distribution is therefore fixed; in other words energy exchange does not take place *via* collisions. Thus, all ions initially generated with insufficient energy ($<E_0$, the hatched area on Fig. 4–8a) to undergo the easiest

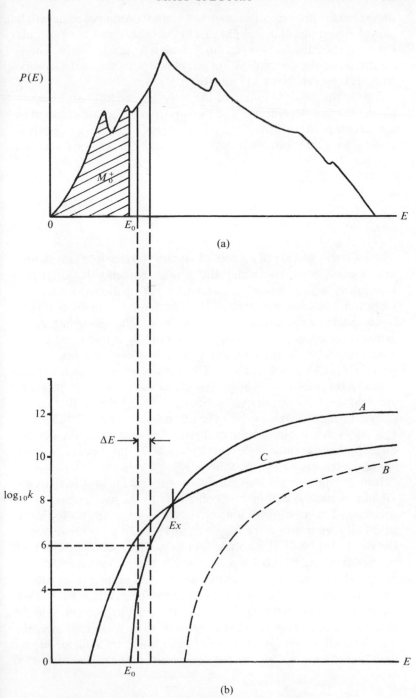

Fig. 4-8

unimolecular decompositions of M^+ cannot decompose and will be recorded as molecular ions (M_0^+). On the other hand, no molecular ion appears in the mass spectrum if there is a zero activation energy for unimolecular decomposition (or, of course, if thermal decomposition is complete prior to ionization).

Ions with energy in excess of the activation energy for decomposition (E_0) may decompose, and the variation of the rate constant (k) for decomposition with internal energy E can be qualitatively understood (but is not quantitatively given) by the classical equation 4–2.

$$k = v\left(\frac{E-E_0}{E}\right)^{s-1} \tag{4-2}$$

where s is the number of degrees of vibrational freedom ($3n-6$ for a non-linear n atom molecule) and v is a frequency factor. For a reaction in which the only requirement is to collect the necessary vibrational activation energy in the reaction co-ordinate (i.e. there is no specific orientation which has to be achieved to attain the activated complex—the analogy in a thermally equilibrated system would be $\Delta S^{\ddagger} = 0$), then, for the hypothetical case, as E approaches ∞, ($E-E_0/E$) approaches 1, and k approaches v. In such circumstances, the molecule would dissociate each time the molecule vibrated along the reaction co-ordinate, and $k_{max} \simeq v \simeq 10^{13}$ sec.$^{-1}$. Curve A (Fig. 4–8b) shows how the rate constant rises with energy to k_{max}; note that the initial rise of k with E is very rapid. As a consequence, energies only slightly in excess of E_0 are needed to bring about decomposition in the source ($\log_{10} k \geqslant 6$).

Ions decomposing within the electrostatic and magnetic analysers are not focused at a point and contribute to the background of the spectrum. However, if an ion m_1 decomposes to an ion m_2 in the field-free region before the magnetic analyser [e.g., in the region of the slit A (Fig. 4–3) of a double focusing mass spectrometer], then the product ion m_2 no longer possesses the normal translational energy eV of a parent or daughter ion formed in the source (due to an ion of charge e falling through an accelerating voltage V). Rather, the translational energy eV of m_1 is partitioned between m_2 and the neutral particle (m_1-m_2) in the ratio of their masses. Hence, the translational energy of m_2 is m_2eV/m_1, lower by a factor m_2/m_1 than that of a normal m_2 formed in the source. The m_2 ion formed in the field-free region does not therefore appear at m_2 on the mass scale but at a lower value m^* given by the relationship:

$$m^* = \frac{m_2}{m_1} m_2 = \frac{m_2^2}{m_1}$$

Such peaks are known as metastable peaks, and normally their presence suggests that the reaction $m_1 \rightarrow m_2$ occurs in one step. Metastable peaks are recognizable because they can occur at non-integral values, and are broader and much less abundant than normal peaks. Metastable peaks can be seen at m/e 133·2 and m/e 123·1 in Fig. 4–5 and indicate that the reactions m/e 185 $\rightarrow$ 157 and m/e 157 $\rightarrow$ 139 are occurring.

The reason for the low abundance of metastable peaks is evident from Fig. 4–8b; metastable peaks are given by reactions with rate constants in the region of 10^4–10^6 sec.$^{-1}$ (contributions from ion lifetimes of about 10 μsec.). The rate constant rises from 10^4–10^6 sec.$^{-1}$ in a very narrow range of energies ΔE (Fig. 4–8a), and so very few ions have these energies.

When two reactions can occur from an ion, they normally compete. So a hypothetical reaction of higher activation energy than that giving rise to curve A is unable to compete effectively and is not observed (curve B). One of lower activation energy and lower frequency factor (curve C) competes; ions of energy $< E_x$ undergo reaction C almost exclusively, while those of energy $> E_x$ undergo reaction A almost exclusively. The physical significance of a low frequency factor (say, in the range 10^8–10^{12} sec.$^{-1}$) is that molecular motions which normally occur in the reactant ion must be 'frozen out' in the transition state; this is analogous to a negative entropy of activation in a system with a Maxwell Boltzmann energy distribution. Thus the theoretical maximum rate is reduced in proportion to the probability of attaining the correct geometry for reaction to occur.

Normally, rearrangement reactions have low frequency factors, and single bond cleavage reactions have high frequency factors. For example, methyl o-toluate competitively loses methanol and a methoxyl radical from its molecular ion. Methanol loss does not occur from the molecular ion of methyl p-toluate and therefore a hydrogen atom of the o-methyl group is involved in the rearrangement.

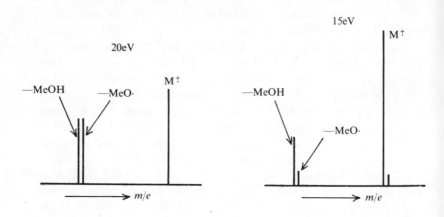

In 20 eV spectra, the $M^+ - 31$ and $M^+ - 32$ ions are of similar abundance, while in spectra obtained at 15 eV, $M^+ - 32$ is much more abundant than $M^+ - 31$ (Fig. 4–9).

Fig. 4–9

In general, the ions formed by primary fragmentation may have enough energy to fragment further, and the same general principles apply to all such reactions occurring in the mass spectrometer. Thus mass spectra result from kinetically controlled reactions, and the relative abundances of ions can easily change by 10 per cent in spectra obtained on different instruments. Large changes in source operating temperatures can lead to even larger changes in relative ion abundances. The following points should be evident from consideration of Fig. 4–8:

(i) Intense molecular ion peaks occur where the easiest unimolecular decomposition of M^+ is a high-energy process (typically $E_0 = 3\text{--}5$ eV). Conversely, weak bonds in molecular ions result in low or negligible M^+ abundances. Table 4–4 serves as a guide for 70 eV spectra of some common classes of compound. 'Strong' implies the molecular ion as the largest peak (base peak) or carrying more than, say, 30 per cent of the total ion current; 'weak' implies a molecular ion of only a few per cent of the abundance of the base peak, and 'medium' applies to intermediate situations.

Table 4–4

Molecular Ion Abundances in Relation to Molecular Structure

Strong	Medium†	Weak or Absent
Aromatic hydrocarbons (ArH)	conjugated olefins	Long-chain aliphatic compounds
ArF	Ar$\{$Br	branched alkanes
ArCl	Ar$\{$I	tertiary aliphatic alcohols
ArCN	ArCO$\{$R	tertiary aliphatic bromides
ArNH$_2$	ArCH$_2\{$R	teriary aliphatic iodides
	ArCH$_2\{$Cl	

† In this column, wavy lines indicate a relatively weak bond.

(ii) When two or more competing reactions occur from a given precursor ion, only the lowest activation energy process gives rise to an abundant metastable peak.

(iii) When two or more decomposition pathways seem feasible from a given molecular ion, then the lowest energy process is observed in the spectrum, and this is probably the *only* observed primary process if the reaction also has a high frequency factor.

4–6. Recognition of Molecular Ions. The Even-Electron Rule

The 'loss' of 14 mass units from a supposed molecular ion should always make one suspect the presence of a homologue, differing in formula by a CH_2-unit. The direct loss of methylene from M^+ is almost never observed, since methylene is such a high energy neutral species. In compounds containing only C, H, O, N, the loss of 5–13 units is virtually impossible, since loss of many hydrogen atoms (or molecules) would have too high an energy requirement. Loss of 3–5 hydrogens is very occasionally observed, and is usually caused by the dehydrogenation of the compound in the inlet system. In such cases, the pattern of ions leading down from M^+ usually shows an ion at each mass (see Fig. 4–10a). The pattern shown in Fig. 4–10b should lead one to think, for example, that A and B are respectively $M^+ - CH_3$ and $M^+ - H_2O$ ions, since the *specific* loss of 3 hydrogens is not found.

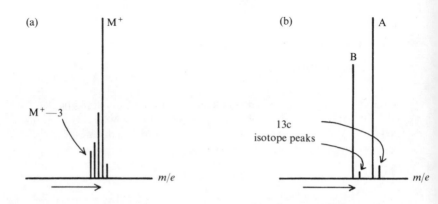

Fig. 4–10

The molecular weights of compounds containing only C, H, (O) are even, as are those of molecules additionally containing an even number of nitrogen atoms. The molecular weights of compounds containing only C, H, N, (O), are odd when the number of nitrogens is odd. Thus when C, H, O compounds lose radicals (CH_3, OCH_3, etc.), odd mass fragment ions result, but when neutral molecules are expelled (H_2O, HCN, olefins, AcOH, etc), even mass fragment ions result.

The even-electron rule states that odd-electron ions decompose by loss of radicals or even-electron molecules, whereas even-electron ions almost always decompose by loss of even-electron molecules. Odd electron ions are those ions which contain an unpaired electron, and include molecular ions (M^+) and fragment ions A^+ formed from M^+ by loss of an even-electron molecule. Even-electron ions normally do not contain unpaired electrons (possess no radical character). To summarize the rule

	M^{+} $\longrightarrow$	B^{+} + radical
or	M^{+} $\longrightarrow$	A^{+} + even-electron molecule
or	A^{+} $\longrightarrow$	C^{+} + radical
or	$A^{!}$ $\longrightarrow$	D^{+} + even-electron molecule
but	E^{+} $\longrightarrow$	F^{+} + even-electron molecule
and not	E^{+} $\longrightarrow$	G^{+} + radical

Exceptions to this rule (i.e., the occurrence of the last reaction listed) are especially uncommon for decompositions which give *metastable peaks*. Exceptions to the rule are found, for example, in some di-iodides, which successively lose iodine radicals from the molecular ion. The physical origin of the rule is presumably associated with the greater inherent stability of most even-electron ions relative to ion-radicals. Since the elimination of a stable even-electron molecule (H_2O, $CH_2{=}CH_2$, HCN, CH_3COOH, etc.) is usually possible, even-electron ions undergo such reactions while preserving their even-electron character; the energy of activation is lowered by partial bond formation (as well as bond stretching) in the transition state. In contrast, odd-electron ions have a choice between eliminating a generally less stable neutral particle (radical loss, no bond formation in the transition state) and forming a generally more stable ion (even-electron), or eliminating a stable even-electron molecule and retaining some instability in the resulting ion-radical.

4–7. Aromatic Compounds

We now apply the principles established in section 4–5 to the mass spectra of aromatic compounds.

The neutral fragments lost when molecular ions in $C_6H_5X^{+}$ decompose upon electron impact are given in Table 4–5; the energetically most facile fragmentation is loss of a methyl radical from the acetophenone molecular ion (requiring only 0·4 *eV*), and the reactions are listed in order of increasing energy requirement from top left to bottom right. For example, reactions such as loss of HF and C_2H_2 from $C_6H_5F^{+}$, and HCN from $C_6H_5CN^{+}$ require several *eV*.

The mass spectrum (Fig. 4–11) of methyl benzoate is fairly typical of that of monosubstituted benzenes, in that the molecular ion is fairly abundant and relatively few fragment ions are observed.

The fragmentation pattern may be rationalized as indicated

Table 4–5

Order of 'Ease of Fragmentation' of some C_6H_5X Compounds

X	Neutral Fragments Lost from $M^{\ddagger}$	X	Neutral Fragments Lost from $M^{\ddagger}$
$COCH_3$	CH_3	OH	CO
$C(CH_3)_3$	CH_3	CH_3	H
$CH(CH_3)_2$	CH_3	Br	Br
CO_2CH_3	OCH_3	NO_2	NO_2, NO
$N(CH_3)_2$	H	NH_2	HCN
CHO	H	Cl	Cl
C_2H_5	CH_3	CN	HCN
OCH_3	CH_2O, CH_3	F	C_2H_2, HF
I	I	H	C_2H_2

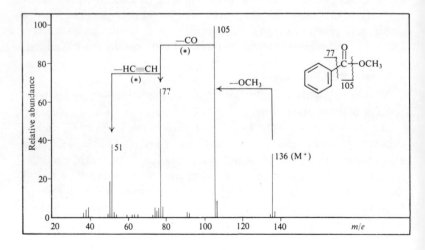

Fig. 4–11

below. If the charge is considered to be localized on the carbonyl oxygen atom, then the homolytic cleavage indicated (by a half-arrow) leads to the benzoyl ion (m/e 105). The m/e 105 ion and its decomposition products are characteristic of benzoyl compounds.†

$$C_6H_5\overset{\overset{\displaystyle +\cdot}{\overset{\displaystyle \ddot{O}}{\|}}}{C}\text{—}OCH_3 \xrightarrow[*]{-OCH_3} C_6H_5C\equiv O^+ \xrightarrow[*]{-CO} C_6H_5^+ \xrightarrow[*]{-C_2H_2} C_4H_3^+$$

$$ m/e\ 105 m/e\ 77 m/e\ 51$$

Benzyl compounds generally afford $C_7H_7^+$ (m/e 91) and its decomposition product $C_5H_5^+$ (m/e 65).

$$C_6H_5CH_2\{R \rceil^{+} \xrightarrow{-\cdot R} C_7H_7^+ \xrightarrow[*]{-C_2H_2} C_5H_5^+$$

$$ m/e\ 91 m/e\ 65$$

If the benzylic bond is very weak (e.g. R = Br), then $C_7H_7^+$ is probably produced as the benzyl ion ($C_6H_5CH_2^+$) at the threshold, but when the benzylic bond is stronger (e.g. R = H), then the tropylium ion may be produced directly.

In the mass spectra of disubstituted benzenes (within the limitations given below), and indeed in the majority of disubstituted aromatic systems, the energetically easier fragmentation (Table 4–5) is in general the only one observed. Thus p-cyano-t-butylbenzene loses only a methyl radical as a primary reaction, and HCN loss from M$^+$ does not occur (Fig. 4–12a). As expected from Table 4–5, a bromine radical is lost exclusively from the molecular ion of p-bromoaniline (Fig. 4–12b), but the table is of limited use when substituent groups fall very close together. For example, the molecular ion of p-chloroaniline competitively loses Cl and HCN (Fig. 4–12c), rather than losing exclusively HCN.

The spectra produced in Figs. 4–12a–c are partial mass spectra, obtained at *low electron beam energies* (this term typically refers to spectra obtained in the 12–20 eV range) so that secondary fragmentations are minimized and multi-step fragmentations avoided. Normally, the most useful information in the mass spectrum comes from the ions in the high mass region, produced by primary and

---

† Here, and subsequently in this chapter, an asterisk under an arrow is used to indicate a decomposition pathway supported by the presence of a metastable peak.

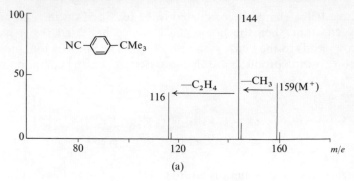

(a)

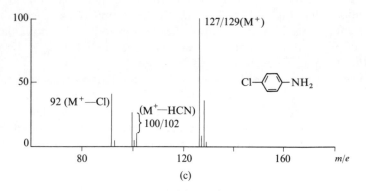

(b)

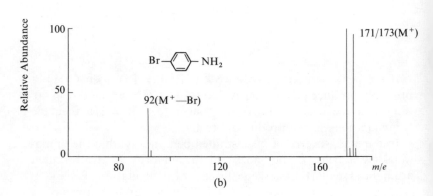

(c)

Fig. 4–12

secondary fragmentation. The secondary fragmentation of m/e 144 in Fig. 4–12a involving loss of ethylene might initially seem surprising, since m/e 144 is most probably correctly represented as c at the instant of its formation. However, it is probable that the energy of activation for the carbonium ion isomerization $c \to d$ (and related fast, reversible carbonium ion rearrangements) is less than that for unimolecular decomposition of c, and that d, once formed, can readily undergo decomposition by loss of ethylene.

Some limitations to the application of Table 4–5 in understanding the mass spectra of di- and multi-substituted aromatic systems are as follows.

(i) *Resonance effects*. Electronic interaction between two substituents may slightly modify the activation energies for reaction relative to ArX. Those substituents $X[N(CH_3)_2, NHCH_3, NH_2, OCH_3, SCH_3, OH$, etc.] which donate electrons to the reaction centre $Y—Z$ enhance the loss of Z as a radical.

Such effects are important only when the two substituents are o- or p-oriented. Thus m-dimethoxybenzene (2, Fig. 4–13) gives major fragment peaks at m/e 108 (M^+—CH_2O) and m/e 109 (M^+—CHO), loss of formaldehyde in one step being a low activation energy process, as in the case of methoxybenzene itself. In contrast, the lowest activation energy process from o-dimethoxybenzene (3, Fig. 4–14) is loss of a methyl radical with formation of the conjugated ion e.

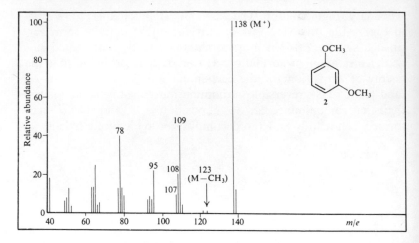

Fig. 4–13

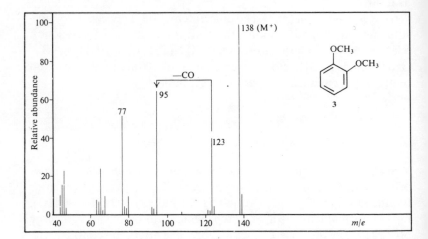

Fig. 4–14

Substituents X (NO$_2$, COR, COOR, etc.), which withdraw electrons from the reaction centre, make the loss of Z as a radical less favourable. If we ignore such interactions, we would expect *p*-methoxytoluene (**5**) to undergo predominantly loss of CH$_2$O and CH$_3$ (Table 4–5), and *p*-nitrotoluene (**4**) to undergo loss of a hydrogen atom (Table 4–5). However, since the pairs of substituents are not widely separated in Table 4–5, resonance effects should additionally be considered. Thus we can expect the energy of activation for the loss of H from *p*-nitrotoluene (**4**) to be raised (NO₂ destabilizes the carbonium ion *f* or its tropylium ion equivalent), and for loss of H from *p*-methoxytoluene (**5**) will be lowered (*g*).

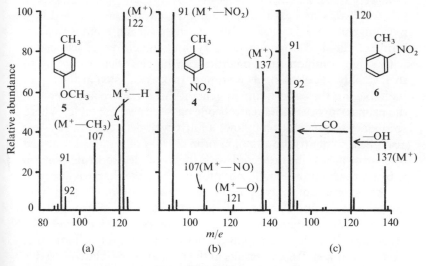

The importance of these considerations is borne out by the high mass regions of the mass spectra of **5** and **4** (Figs. 4–15a, b); M⁺—H is the most abundant daughter ion arising from primary decomposition of **5**, but the loss of H from *p*-nitrotoluene (**4**) is negligible relative to NO and NO₂ loss.

(ii) *Ortho effects.* When two substituents are *meta* or *para* they cannot interact through space, but if *ortho* they can, so that a neutral particle (whose constituents originate in part from each substituent)

Fig. 4–15

may be eliminated with a relatively low activation energy. For example, the presence of a methyl group *ortho* to the nitro-function in *o*-nitrotoluene (**6**) introduces a new, low energy process, namely loss of a hydroxyl radical to give *m/e* 120, perhaps with formation of the bicyclic ion shown (contrast Figs. 4–15c and 4–15a).

m/e 120   **7**, X = O, S, NH, CH$_2$ etc.

One of the most useful ortho-effects in structure elucidation is associated with ROH loss (in competition with OR loss) from various kinds of aromatic esters carrying X—H groups in the *ortho* position (**7**).

4–8. Aliphatic Compounds

A. *Primary Single Bond Cleavages*

As we have already seen, primary fragmentation will normally be largely associated with cleavage of the weakest bond in the molecular ion. If this process is a single bond cleavage (as opposed to a rearrangement), it is usually obvious why a certain bond is weak in terms of classical organic chemistry, and the cleavages associated with some common functionalities can be classified according to their facility (Table 4–6). The primary processes listed at the top of the table are the easiest, and in general the energy requirements of the primary processes increase as one passes down the table. Where secondary fragmentation follows a fairly well-defined pathway, it is also given. Note (i) the common primary loss of a radical followed by loss of an olefin, in accord with the even-electron rule (section 4–6), (ii) the usefulness of the notation in which the charge is localized on the heteroatom (corresponding to removal of an electron from the highest occupied molecular orbital). Cleavage of groups from the carbon atom adjacent to the heteroatom (N, O, S) is energetically favourable; it is represented as a homolytic process ($\frown$, indicating a one-electron shift), and results in the formation of relatively stable immonium, oxonium ions, etc.

Table 4–6

Primary Single Bond Cleavage Processes associated with some Common Functional Groups

Functional Group	Fragmentation
Amine	$R_2(CH_2)_n\overset{+\cdot}{N}\overset{\frown}{-}CH_2-R_3 \xrightarrow{-R_3\cdot}$ $R_2(CH_2)_n\overset{+}{N}=CH_2 \xrightarrow{-\text{olefin}} H\overset{+}{N}=CH_2$ with R_1 substituents
Ketal	$\underset{R_2}{\overset{R_1}{>}}C\overset{\overset{\cdot+}{O}}{\underset{O}{<}} \xrightarrow{-R_2\cdot} R_1-C\overset{\overset{+}{O}}{\underset{O}{<}}$
Iodide	$R{-}I\rceil^{+} \xrightarrow{-I\cdot} R^+$
Ether (X = O) Thioether (X = S)	$R_2(CH_2)_n\overset{+\cdot}{X}{-}CH{-}R_1 \xrightarrow{-R_3\cdot}$ with R_3 $R(CH_2)_n\overset{+}{X}=CH-R_1 \xrightarrow{-\text{olefin}} H\overset{+}{X}=CHR_1$
Ketone	$\underset{R_2}{\overset{R_1}{>}}C=O^{+} \xrightarrow{-R_2\cdot} R_1-C\overset{+}{\equiv}O \xrightarrow{-CO} R_1^+$
Alcohol (X = O) Thiol (X = S)	$R_1{-}CH_2{-}XH \xrightarrow{-R_2\cdot} R_1CH=\overset{+}{X}H$ with $\overset{\cdot+}{R_2}$
Bromide	$RBr\rceil^{+} \xrightarrow{-Br\cdot} R^+$
Ester	$R_1\overset{\overset{+\cdot}{O}}{\underset{\|}{C}}OR_2 \longrightarrow \overset{+}{O}\equiv C{-}OR_2$ $\longrightarrow R_1C\equiv O^+$

When primary fragmentation can lead to competition between losses of two or more different radicals, in 70 eV spectra the loss of the larger radical is usually dominant. Other things being equal, the ease with which hydrocarbon radicals are lost is in the order tertiary > secondary > primary.

It is very useful to remember the m/e values associated with the simplest members of the ion types produced by some of the fragmentations listed in Table 4–6. If the groups attached to these 'simplest ion types' are saturated hydrocarbon groups, then the m/e values fall in the series m/e $(x + 14n)$, where $(m/e)x$ is the first member of the series (Table 4–7). This relation arises because of the addition of integral numbers of CH_2 groups.

Table 4–7

Useful Ion Series

Functional Group	Simplest Ion Type	Ion Series (m/e)
Amine	$CH_2{=}\overset{+}{N}H_2$ m/e 30	30, 44, 58, 72, 86, 100, . . .
Ether } Alcohol }	$CH_2{=}\overset{+}{O}H$ m/e 31	31, 45, 59, 73, 87, 101, . . .
Ketone	$CH_3C{\equiv}\overset{+}{O}$ m/e 43	43, 57, 71, 85, 99, 113, . . .
[Hydrocarbon]	$C_2H_5^+$ m/e 29	29, 43, 57, 71, 85, 99, 113, . . .

B. Rearrangement Reactions

The elimination of neutral molecules directly from the molecular ions of aliphatic secondary and tertiary amines, ketals, iodides and ethers is not normally observed; the primary processes of radical loss (Table 4–6) are too favourable. However, in nearly all aliphatic carbonyl compounds, a primary rearrangement fragmentation is observed, in which a γ-hydrogen, if available, migrates to the carbonyl oxygen. Usually, a charged enol is formed with elimination of a neutral olefin. The m/e values of the ions formed from various carbonyl compounds are given in Table 4–8.

Table 4–8

m/e **Values of Some Rearrangement Ions Found in the Mass Spectra of Carbonyl Compounds**

$$\left[\begin{array}{c} R_1 \diagdown \quad H \\ CH \nearrow \quad O \\ | \leftarrow \quad \parallel \\ CH \diagdown \; C \\ R_2 \diagup \quad CH_2 \diagdown X \end{array}\right]^{\ddagger} \xrightarrow[\left(\substack{R_1 \text{ or } R_2 \\ = \text{alkyl or } H}\right)]{- R_1CH=CHR_2} \left[\begin{array}{c} OH \\ | \\ C \\ \diagup \diagdown \\ CH_2 \quad X \end{array}\right]^{\ddagger}$$

Compound	X	*m/e*
Aldehyde	H	44
Ketone (methyl)	CH_3	58
Ketone (ethyl)	C_2H_5	72
Acid	OH	60
Ester (methyl)	OCH_3	74
Ester (ethyl)	OC_2H_5	88
Amide	NH_2	59

Another useful group of rearrangement reactions consists of those associated with the loss of small stable neutral molecules from molecular ions in which primary radical loss is not particularly favourable. These include loss of H_2O (18 m.u.) from many alcohols, HF (20 m.u.) from many fluorides and CH_3COOH (60 m.u.) from many acetates.

The loss of a methoxyl radical from methyl esters (to give $M^+ - 31$) is quite characteristic, but larger alcohol groups in esters (ethyl, propyl, butyl, etc.), in addition to losing the alkoxyl group, also undergo single and double hydrogen rearrangements to give ionized carboxylic acid and protonated carboxylic acid species, respectively:

$$\left[\begin{array}{c} O \\ \parallel \\ R_1COR \end{array}\right]^{\ddagger} \underset{\substack{-(R-2H) \\ \searrow}}{\overset{\substack{-(R-H) \\ \nearrow}}{}} \begin{array}{l} \left[R_1C\diagup^{O}_{\diagdown OH}\right]^{\ddagger} \\[2ex] R_1C\diagup^{\overset{+}{O}H}_{\diagdown OH} \end{array}$$

The proportion of double hydrogen rearrangement increases with increasing size of R.

It is noteworthy that isomeric primary, secondary and tertiary saturated aliphatic carbonium ions normally interconvert at rates which are fast compared to their unimolecular decomposition. The characteristic decomposition pathways (Table 4–9) of carbonium ions are by hydrogen-rearrangement fragmentations, in accord with the even-electron rule (section 4–6).

Table 4–9

Decomposition Pathways of Some Saturated Carbonium Ions

Precursor Ion	Decomposition Products	Transition (m/e)	m^*
$C_2H_5^+$	$C_2H_3^+ + H_2$	$29 \longrightarrow 27$	25·14
$C_3H_7^+$	$C_3H_5^+ + H_2$	$43 \longrightarrow 41$	39·09
$C_4H_9^+$	$C_3H_5^+ + CH_4$	$57 \longrightarrow 41$	29·49
$C_5H_{11}^+$	$C_3H_7^+ + C_2H_4$	$71 \longrightarrow 43$	26·04
$C_6H_{13}^+$	$C_4H_9^+ + C_2H_4$ and	$85 \longrightarrow 57$	38·22
	$C_3H_7^+ + C_3H_6$	$85 \longrightarrow 43$	21·75
$C_7H_{15}^+$	$C_4H_9^+ + C_3H_6$	$99 \longrightarrow 57$	32·82

Some of the points made so far in this section are now illustrated by the 70 eV spectra of some hydrocarbons and some monofunctional compounds.

(a) *Hydrocarbons*

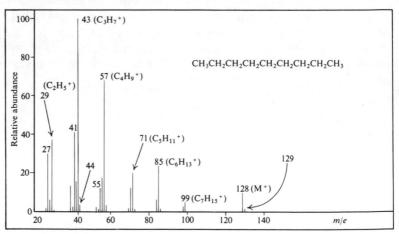

Fig. 4–16

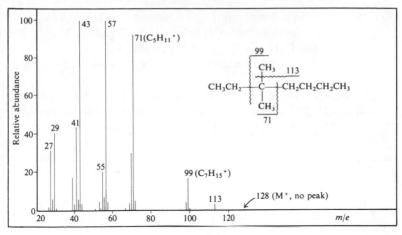

Fig. 4–17

Points: (i) Hydrocarbon ion series m/e 29, 43, 57, etc. (Table 4–7).

(ii) Very facile fragmentation in branched hydrocarbon (Fig. 4–17) and therefore no M^+ peak; formation of tertiary carbonium ions favoured (m/e 113, 99, 71) in primary fragmentation.

(iii) Ion series $C_nH^+_{2n-1}$ (m/e 27, 41, 55, . . .) of moderate abundance at low mass end of spectrum (Table 4–9).

(iv) Low mass ions (m/e 41, 43, 57) very abundant in 70 eV spectra.

(b) *Ketones*

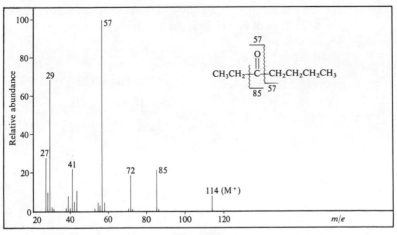

Fig. 4–18

Points: (i) Loss of alkyl groups attached to the carbonyl function (Table 4–6) leads to acylium ions $C_4H_9C\equiv O^+$ (m/e 85) and $C_2H_5C\equiv O^+$ (m/e 57) (Table 4–7).

    (ii) Rearrangement ion at m/e 72 suggests ethyl ketone (Table 4–8).

    (iii) m/e 57 shown by high resolution to be due to both $C_4H_9^+$ and $C_2H_5C\equiv O^+$ (saturated carbonium and acyl ions give same m/e values—Table 4–7).

(c) *Ethers*

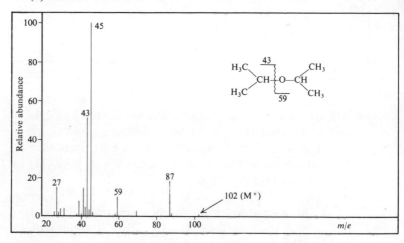

Fig. 4–19

Points: (i) Ions at m/e 45, 59, 87 suggest an ether or alcohol (Table 4–7); lack of H_2O loss from M^+ indicates more probably an ether.

    (ii) Reactions occurring are

m/e 43 and 59 formed as indicated in Fig. 4–19.

(d) *Amines*

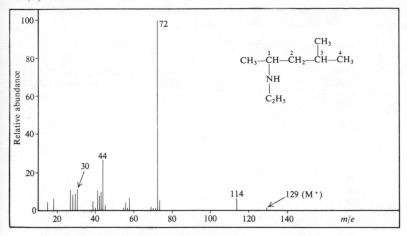

Fig. 4–20

Points: (i) Odd molecular weight and ions at m/e 30, 44, 72, and 114 suggest a saturated amine (Table 4–7).

     (ii) Largest alkyl radical preferentially lost in decomposition of M^+.

$$CH_3CH \overset{.+}{-} \overset{|}{N}C_2H_5 \longrightarrow CH_3CH{=}\overset{+}{N}\overset{|}{C}H_2 \xrightarrow{-C_2H_4} CH_3CH{=}\overset{+}{N}H_2$$

M^+, m/e 129 m/e 72 m/e 44

C. *Polyfunctional Molecules*

It has already been pointed out that the energy requirements for decomposition generally increase as one progresses down the functional groups listed in Table 4–6. By examining the relative fractions of ion current (i_{OCH_3} and i_X) arising from cleavages initiated by CH_3O and X groups in the bifunctional decanes (**8**) at 70 eV, it is possible to place some functional groups in an order in terms of their ability to control fragmentation in the mass spectra of polyfunctional molecules (Table 4–10).

$$C_2H_5\underset{\overset{|}{X}}{CH}{-}(CH_2)_4{-}\underset{\overset{|}{OCH_3}}{CH}C_2H_5$$

8

Table 4–10

Relative Propensities of Common Functional Groups to Produce Cleavage in Compounds of the General Formula 8

X	i_x†	X	i_x†
—COOH	2	—SCH$_3$	90
—CH$_2$OH	6	—OCH$_3$	100
—Cl	8	—NHCOCH$_3$	121
—COOCH$_3$	17	—I	131
—Br	23	$\underset{\diagdown}{\diagup}\text{C}\underset{\text{O—CH}_2}{\overset{\text{O—CH}_2‡}{\diagup}}$	514
—OH	25		
—SH	36	—NH$_2$	616
$\overset{\diagdown}{\underset{\diagup}{}}$C=O‡	41	—N(CH$_3$)$_2$	1200

† Relative to i_{OCH_3} arbitrarily taken as 100 units.
‡ Data refer to the 3-methoxy-8-ketodecane and 3-methoxy-8-keto-decane ethylene ketal derived from **8**.

The trends may be discerned by reference to the mass spectra (Figs. 4–21 and 4–22) of 8-methylmercapto-3-methoxydecane (**9**) and 3-methoxydecan-8-one ethylene ketal (**10**). Note that in the former case the ether and thioether control initial fragmentation to similar extents in producing m/e 73 (h) and m/e 89 (i) in similar

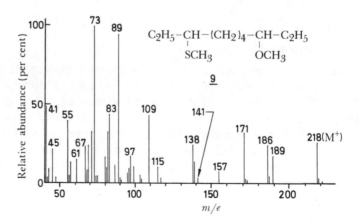

Fig. 4–21

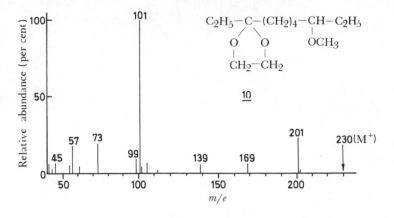

Fig. 4-22

abundance. The detailed decomposition pathways are summarized in the reaction scheme.

$$C_2H_5CH\overset{+}{=}OH \qquad\qquad C_2H_5CH\overset{+}{=}SH$$
$$h,\ m/e\ 73 \qquad\qquad\qquad i,\ m/e\ 89$$

$$\left[C_2H_5-CH-(CH_2)_4-CH-C_2H_5\right]^{\ddagger}$$
$$\qquad\quad SCH_3 \qquad\qquad OCH_3$$

9

$-C_2H_5$ $-CH_3OH$ $-SCH_3$

m/e 189 m/e 186 m/e 171

$-CH_3OH$ $-CH_3SH$ $-CH_3SH$ $-CH_3OH$

m/e 157 m/e 141 m/e 138 m/e 139

In contrast, in the mass spectrum of **10**, the activation energy for formation of the highly stabilized oxonium ion m/e 101 is so small that formation of m/e 73, and other primary processes, compete relatively poorly.

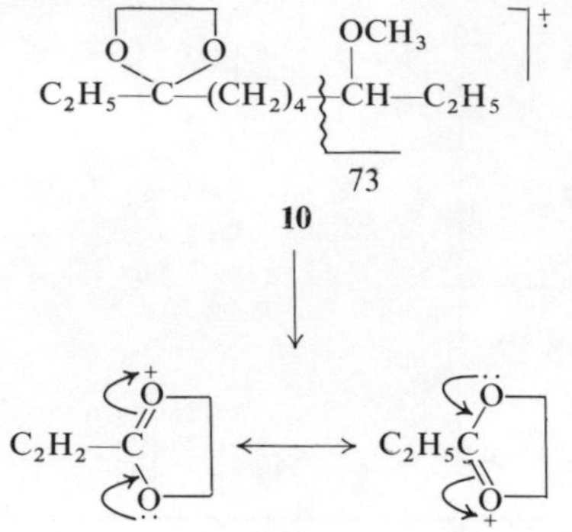

$$C_2H_5\overset{O}{\underset{}{C}}(CH_2)_4\left\{\overset{OCH_3}{\underset{}{CH}}-C_2H_5\right]^{\ddagger}$$

73

10

$$C_2H_2-\overset{+O}{\underset{O}{C}} \longleftrightarrow C_2H_5\overset{\ddot{O}}{\underset{+O}{C}}$$

m/e 101

4–9. Stephenson's Rule

Some observations made many years ago on the mass spectra of alkanes apply to fragmentations in general: for the reactions:

$$AB^{\ddagger} \longrightarrow A^+ + B\cdot$$

or

$$AB^{\ddagger} \longrightarrow A\cdot + B^+$$

the former occurs predominantly or exclusively if I.P.(A·) < I.P.(B·). Similarly for the reactions:

$$CD^{\ddagger} \longrightarrow C^{\ddagger} + D$$

or

$$CD^{\ddagger} \longrightarrow C + D^{\ddagger}$$

the former occurs predominantly or exclusively if I.P.(C) < I.P.(D).

The rule tells us which portion retains the charge if a given type of fragmentation occurs, and it is a simple matter to have the 'feel' of relative ionization potentials if we remember that extending the conjugation of a system generally lowers its ionization potential. Thus, in the cyclohexenyl systems **11** and **12**, which typically undergo

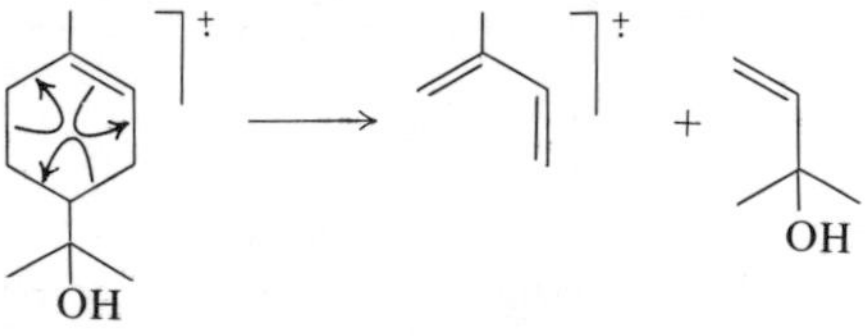

11

m/e 68
I.P. = 9·0 eV   I.P. ≃ 9·4 eV

12   *m/e* 86
                    I.P. = 8·7 eV   I.P. = 9·0 eV

a retro-Diels-Alder fragmentation on electron impact, the diene system appears as the charged fragment in the spectrum of **11**, but the ene portion retains the charge in the spectrum of **12**.

The following relative I.P.s and their consequences are also noteworthy.

                    m/e 58
                I.P. = 8·8 eV   I.P. = 9·8 eV

                m/e 104
            I.P. = 8·5 eV I.P. = 8·8 eV

4–10. Interpreting the Spectrum of an Unknown

(*a*) *Counting the spectrum; impurities.* In many cases, convenient starting points for counting the spectrum are the peaks at *m/e* 28 (N_2^+) and *m/e* 32 (O_2^+). If a *background spectrum* is run before a sample is introduced, weak peaks are often present at *m/e* 41, 43, 55 and 57 (hydrocarbon background). Samples which have been extensively 'handled' (e.g. on thin-layer plates, columns, and greased apparatus) may contain impurity peaks (Table 4–11).

(*b*) *The Molecular ion.* Check whether the peak at highest mass is likely to be a molecular ion (section 4–6). Check whether the peaks

Table 4–11

Some Common Impurity Peaks

m/*e* Values	Cause
149, 167, 279	Plasticizers (phthalic acid derivatives)
129, 185, 259, 329	Plasticizer (tri-n-butyl acetyl-citrate)
133, 207, 281, 355, 429	Silicone grease
99, 155, 211	Plasticizer (tributyl phosphate)

immediately below an assumed molecular ion correspond to the loss of plausible neutral particles (Table 4–12). Infer nature of bonds in molecular ion (e.g., all strong, some weak) from molecular ion abundance (e.g., Table 4–4). Note whether molecular weight is odd or even (section 4–6), and any characteristic isotope patterns (section 4–4).

(*c*) *Fragmentation pattern.* Use mass differences from M^+ (Table 4–12) and table of common *m*/*e* values (Table 4–13) to give a preliminary indication of which functional groups may be present, and partial structural information. Assign any metastable peaks. Check on characteristic ion series (Table 4–7). Ensure in the case of polyfunctional molecules that the structural conclusions are consistent not only with the expected fragmentation (e.g., Tables 4–5 and 4–6) but with expected competition between fragmentation of the functionalities (Tables 4–5 and 4–10).

It must be remembered that gas phase unimolecular decompositions are inevitably more varied in a wide range of compounds than can be conveyed in a few pages of text. Since spectroscopic transitions are in general more predictable than the precise course of chemical reactions, the rules for interpreting UV, IR and NMR spectra are necessarily more precise than those for interpreting the mass spectrum. The cardinal rule to remember is that the precursor ions break to give the energetically most favourable combinations of ion and radical (or ion and neutral molecule).

4–11. Isotopic Labelling and High Resolution; the Shift Technique and Skeletal Rearrangement

Frequently, the fragmentation pattern of a molecule is unambiguous because the atomic constituents can only be combined in one manner to give the observed *m*/*e* values. However, in more

Table 4–12

Some Common Losses from Molecular Ions

Ion	Groups Commonly Associated with the Mass Lost	*Possible* Inference
M − 1	H	—
M − 2	H_2	—
M − 14	—	Homologue?
M − 15	CH_3	—
M − 16	O	Ar—NO_2, $\geqslant \overset{+}{N}$—$\overset{-}{O}$, sulphoxide
M − 16	NH_2	$ArSO_2NH_2$, —$CONH_2$
M − 17	OH	—
M − 17	NH_3	—
M − 18	H_2O	Alcohol, aldehyde, ketone, etc.
M − 19	F	⎫ Fluorides
M − 20	HF	⎭
M − 26	C_2H_2	Aromatic hydrocarbon
M − 27	HCN	⎰ Aromatic nitriles ⎱ Nitrogen heterocycles
M − 28	CO	Quinones
M − 28	C_2H_4	⎰ Aromatic ethyl ethers ⎱ Ethyl esters, n-propyl ketones
M − 29	CHO	—
M − 29	C_2H_5	Ethyl ketones, Ar—n—C_3H_7
M − 30	C_2H_6	—
M − 30	CH_2O	Aromatic methyl ether
M − 30	NO	Ar—NO_2
M − 31	OCH_3	Methyl ester
M − 32	CH_3OH	Methyl ester
M − 32	S	—
M − 33	$H_2O + CH_3$	—
M − 33	HS	⎫ Thiols
M − 34	H_2S	⎭
M − 41	C_3H_5	Propyl ester
M − 42	CH_2CO	⎰ Methyl ketone ⎱ Aromatic acetate, $ArNHCOCH_3$
M − 42	C_3H_6	⎰ n- or iso-butyl ketone, ⎱ Aromatic propyl ether, Ar—n—C_4H_9
M − 43	C_3H_7	Propyl ketone, Ar—n—C_4H_9
M − 43	CH_3CO	Methyl ketone
M − 44	CO_2	⎰ Ester (skel. rearr.) ⎱ Anhydride
M − 44	C_3H_8	—
M − 45	CO_2H	Carboxylic acid
M − 45	OC_2H_5	Ethyl ester
M − 46	C_2H_5OH	Ethyl ester
M − 46	NO_2	Ar—NO_2
M − 48	SO	Aromatic sulphoxide
M − 55	C_4H_7	Butyl ester
M − 56	C_4H_8	⎰ Ar—n-C_5H_{11}, ArO—n-C_4H_9 ⎨ Ar—iso-C_5H_{11}, ArO—iso-C_4H_9 ⎱ Pentyl ketone
M − 57	C_4H_9	Butyl ketone
M − 57	C_2H_5CO	Ethyl ketone
M − 58	C_4H_{10}	—
M − 60	CH_3COOH	Acetate

Table 4–13

Masses and Some Possible Compositions of Common Fragment Ions

m/e	Groups Commonly Associated with the Mass	*Possible* Inference
15	CH_3^+	—
18	H_2O^+	—
26	$C_2H_2^+$	—
27	$C_2H_3^+$	—
28	CO^+, $C_2H_4^+$, N_2^+	—
29	CHO^+, $C_2H_5^+$	—
30	$CH_2{=}\overset{+}{N}H_2$	Primary amine?
31	$CH_2{=}\overset{+}{O}H$	Primary alcohol?
36/38(3:1)	HCl^+	—
39	$C_3H_3^+$	—
40†	$Argon^+$, $C_3H_4^+$	—
41	$C_3H_5^+$	—
42	$C_2H_2O^+$, $C_3H_6^+$	—
43	CH_3CO^+	CH_3COX
43	$C_3H_7^+$	C_3H_7X
44	$C_2H_6N^+$	Some aliphatic amines
44	$O{=}C{=}\overset{+}{N}H_2$	Primary amides
44	CO_2^+, $C_3H_8^+$	—
44	$CH_2{=}CH(OH)^+$	Some aldehydes
45	$CH_2{=}\overset{+}{O}CH_3$	⎫
	$CH_3CH{=}\overset{+}{O}H$	⎬ Some ethers and alcohols
47	$CH_2{=}\overset{+}{S}H$	Aliphatic thiol
49/51(3:1)	CH_2Cl^+	—
50	$C_4H_2^+$	Aromatic compound
51	$C_4H_3^+$	C_6H_5X
55	$C_4H_7^+$	—
56	$C_4H_8^+$	—
57	$C_4H_9^+$	C_4H_9X
57	$C_2H_5CO^+$	⎰ Ethyl ketone ⎱ Propionate ester
58	$CH_2{=}C(OH)CH_3^+$	⎰ Some methyl ketones ⎱ Some dialkyl ketones
58	$C_3H_8N^+$	Some aliphatic amines
59	$COOCH_3^+$	Methyl ester
59	$CH_2{=}C(OH)NH_2^+$	Some primary amides
59	$C_2H_5CH{=}\overset{+}{O}H$	$C_2H_5CH(OH){-}X$
59	$CH_2{=}\overset{+}{O}{-}C_2H_5$ and isomers	Some ethers
60	$CH_2{=}C(OH)OH^+$	Some carboxylic acids
61	$CH_3CO(OH_2)^+$	$CH_3COOC_nH_{2n+1}(n > 1)$

† Appears as a doublet in the presence of argon from air; useful as a reference point in counting the mass spectrum.

Table 4–13 *continued*

m/e	Groups Commonly Associated with the Mass	*Possible* Inference
61	$CH_2CH_2SH^+$	Aliphatic thiol
66	$H_2S_2^+$	Dialkyl disulphide
69	CF_3^+	—
68	$CH_2CH_2CH_2CN^+$	—
69	$C_5H_9^+$	—
70	$C_5H_{10}^+$	—
71	$C_5H_{11}^+$	$C_5H_{11}X$
71	$C_3H_7CO^+$	$\left\{\begin{array}{l}\text{Propyl ketone}\\\text{Butyrate ester}\end{array}\right.$
72	$CH_2{=}C(OH)C_2H_5^+$	Some ethyl alkyl ketones
72	$C_3H_7CH{=}\overset{+}{N}H_2$ and isomers	Some amines
73	$C_4H_9O^+$	—
73	$COOC_2H_5^+$	Ethyl ester
73	$(CH_3)_3Si^+$	$(CH_3)_3SiX$
74	$CH_2{=}C(OH)OCH_3^+$	Some methyl esters
75	$(CH_3)_2Si{=}\overset{+}{O}H$	$(CH_3)_3SiOX$
75	$C_2H_5CO(OH_2)^+$	$C_2H_5COOC_nH_{2n+1}(n>1)$
76	$C_6H_4^+$	$\left\{\begin{array}{l}C_6H_5X\\XC_6H_4Y\end{array}\right.$
77	$C_6H_5^+$	C_6H_5X
78	$C_6H_6^+$	C_6H_5X
79	$C_6H_7^+$	C_6H_5X
79/81 (1:1)	Br^+	—
80/82 (1:1)	HBr^+	—
80	$C_5H_6N^+$	
81	$C_5H_5O^+$	
83/85/87 (9:6:1)	$HCCl_2^+$	$CHCl_3$
85	$C_6H_{13}^+$	$C_6H_{13}X$
85	$C_4H_9CO^+$	C_4H_9COX
85		
85		
86	$CH_2{=}C(OH)C_3H_7^+$	Some propyl alkyl ketones
86	$C_4H_9CH{=}\overset{+}{N}H_2$ and isomers	Some amines

Table 4–13 *continued*

m/e	Groups Commonly Associated with the Mass	*Possible* Inference
87	$CH_2{=}CH{-}\overset{+OH}{\overset{\|}{C}}{-}OCH_3$	$XCH_2CH_2COOCH_3$
91	$C_7H_7^+$	$C_6H_5CH_2X$
92	$C_7H_8^{+\cdot}$	$C_6H_5CH_2$alkyl
92	$C_6H_6N^+$	(pyridyl)$-CH_2X$
91/93 (3:1)	(cyclopentyl chloride cation)	n-Alkyl chloride ($\geqslant$ hexyl)
93/95 (1:1)	CH_2Br^+	—
94	$C_6H_6O^{+\cdot}$	C_6H_5O-alkyl (alkyl $\neq CH_3$)
94	(pyrrolyl)$-C{\equiv}\overset{+}{O}$	(pyrrolyl)$-COX$
95	(furyl)$-C{\equiv}\overset{+}{O}$	(furyl)$-COX$
95	$C_6H_7O^+$	CH_3-(furyl)$-CH_2X$
97	$C_5H_5S^+$	(thienyl)$-CH_2X$
99	(vinyl dioxolane cation)	(dimethyl dioxaspiro structure)
99	(dihydropyranone cation)	$X-$(tetrahydropyranone)
105	$C_6H_5CO^+$	C_6H_5COX
105	$C_8H_9^+$	$CH_3-C_6H_4CH_2X$
106	$C_7H_8N^+$	CH_3-(pyridyl)$-CH_2X$

Table 4–13 *continued*

m/e	Groups Commonly Associated with the Mass	*Possible* inference
107	$C_7H_7O^+$	
107/109 (1:1)	$C_2H_4Br^+$	—
111		
121	$C_8H_9O^+$	
122	C_6H_5COOH	} Alkyl benzoates
123	$C_6H_5COOH_2^+$	
127	I^+	—
128	HI^+	—
135/137 (1:1)		n-Alkyl bromide ($\geqslant$ hexyl)
130	$C_9H_8N^+$	
141	CH_2I^+	—
147	$(CH_3)_2Si{=}\overset{+}{O}{-}Si(CH_3)_3$	—
149		Dialkyl phthalate
160	$C_{10}H_{10}NO^+$	
190	$C_{11}H_{12}NO_2^+$	

complicated molecules, particularly those containing oxygen and nitrogen in addition to carbon and hydrogen, a given m/e value can often arise in many ways. Thus in a C-, H-, O- and N-containing molecule $M-28$ ions could correspond to $M-H_2CN$, $M-CO$ or $M-C_2H_4$, whereas $M-27$ could plausibly be due to $M-C_2H_3$ or $M-HCN$. In these cases isotopic labelling or high resolution measurements must be employed to uncover the fragmentation mechanism or to aid in structure elucidation.

The most common isotope employed in isotopic labelling work is deuterium. Some useful methods for introducing deuterium into organic molecules are summarized in reactions and exchange processes shown below.

1. $-CH_2-CH=CH-CH_2-\xrightarrow[Pd/C]{D_2} -CH_2-CDH-CDH-CH_2-$

2. $-CH_2-\overset{\overset{O}{\|}}{C}-CH_2-\xrightarrow[D_2O/CH_3OD]{NaOD} -CD_2-\overset{\overset{O}{\|}}{C}-CD_2-$

3. $-CH_2-\overset{\overset{O}{\|}}{C}-CH_2-\xrightarrow{LiAlD_4} -CH_2-\overset{\overset{OH}{|}}{\underset{\underset{D}{|}}{C}}-CH_2-$

4. $-CH_2-\overset{\overset{OTs}{|}}{CH}-CH_2-\xrightarrow{LiAlD_4} -CH_2-CHD-CH_2-$

5. $ROH, RNH_2, RCOOH \xrightarrow{D_2O} ROD, RND_2, RCOOD$

6. $RCOOH \xrightarrow[CD_2N_2]{D_2O} RCOOCD_3$

The exchange of active hydrogen for deuterium (process 5) is extremely convenient because it can be carried out by introducing deuterium oxide into the heated inlet system of the spectrometer concurrently with the sample. The preparation of labelled methyl esters (reaction 6) does not actually require the preparation of labelled diazomethane, since the hydrogens of CH_2N_2 are partially exchanged for deuterium in the presence of acid and D_2O. Active methylene compounds such as acetylacetone (13) are in equilibrium with their enolates, and so if 13 is introduced into the spectrometer with D_2O, the spectrum of the d_2-derivative 14 is obtained. Surprisingly, the spectrum of 13 contains an ion at m/e 72 ($M-28$), which can only be formed in a rearrangement reaction. In the spectrum of 14, the m/e 72 peak of the parent compound is shifted to m/e 74, indicating that both deuterium atoms are retained in the M $-$ 28 ion. This observation strongly suggests that the M $-$ 28 peak is formed by elimination of carbon monoxide rather than ethylene (which could in principle be expelled from a rearranged molecular ion) and this supposition is confirmed by exact mass measurements on m/e 72 ($C_4H_8O^+$ and not $C_3H_4O_2^+$) using a double focusing mass spectrometer. Such processes (where a fragment $AC^{\ddagger}$, not present in the original molecule ABC, is generated in the mass spectrometer) are known as skeletal rearrangements and occur in perhaps 5–10 per cent of organic compounds. They most usually occur when a molecular ion contains no one bond that is particularly weak, but elimination of a stable neutral molecule from within the ion is energetically favourable. Unrecognized skeletal rearrangements could, in the absence of other evidence to the contrary, lead one to erroneous structural conclusions.

$$CH_3COCH_2COCH_3 \xrightarrow{D_2O} CH_3COCD_2COCH_3$$

13, M^+ at m/e 100   14, M^+ at m/e 102

Carbonyl groups may be labelled with ^{18}O and nitrogen compounds with ^{15}N.

The structure elucidation of natural products by mass spectrometry has been greatly assisted by the *shift technique*. The shift technique has been employed chiefly for the structure elucidation of some classes of alkaloids, for which it can be shown that the modes of fragmentation are the same for a given basic skeleton. A classic case is provided by the mass spectrum of deacetylaspidospermine (15) which exhibits the base peak at m/e 124 and prominent ions at m/e 160, m/e 174 and m/e 284 (M $-$ 28). The m/e 284 ion j is formed by elimination of ethylene from the molecular ion, the driving force being aromatization of ring B. Allylic cleavage in j then gives the immonium ion l (m/e 124) and the indolylmethyl radical k, both of which are relatively stable fragments. If the allylic cleavage occurs with charge retention by the indole moiety, then m/e 160 (m) results; m/e 174 (n) is the higher homologue of m/e 160.

Provided that these fragmentation processes are predominantly independent of the nature of substituents on the basic skeleton, the approximate location of various substituents in related natural products can therefore be inferred from the shifts of the characteristic ions j to n. For example, demethoxydeacetylaspidospermine (16) gives ions analogous to j, m and n at m/e 254, 130 and 144

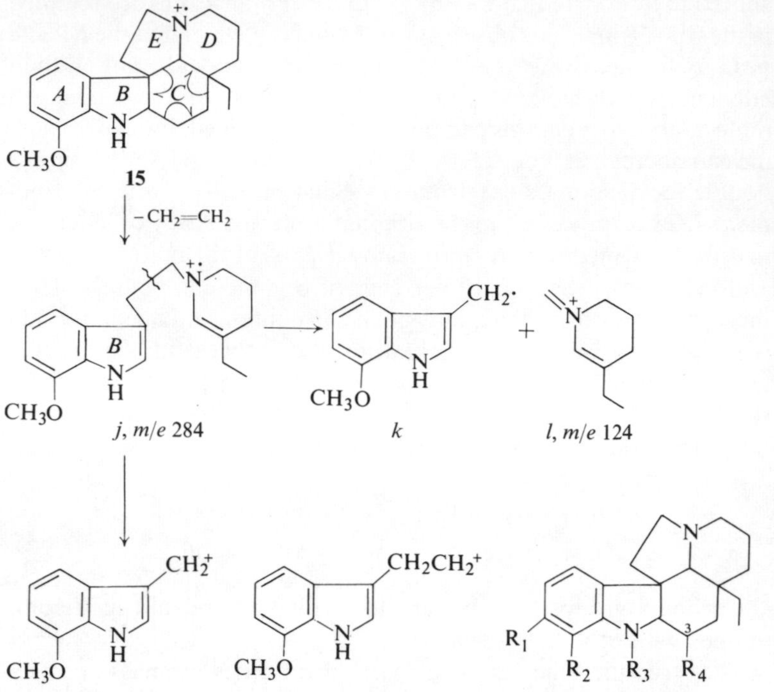

15

$-CH_2=CH_2$

j, m/e 284 k l, m/e 124

m, m/e 160 n, m/e 174

16, $R_1 = R_2 = R_3 = R_4 = H$
17, $R_1 = R_2 = OCH_3$,
    $R_3 = COCH_3$, $R_4 = OH$

respectively, while l remains at m/e 124. In contrast, the spectrum of spegazzinidine dimethyl ether (17) contains an M − 44 ion due to the elimination of the C—3 hydroxyl function in the primary fragmentation process. The base peak is at m/e 124 (l), but analogues of m and n occur at m/e 190 and m/e 204; the shifts of 30 mass units correspond to an additional aromatic methoxyl function, since the N-acetyl group is eliminated as ketene (see M − 42, Table 4–12) during the fragmentation process. This analysis therefore serves to associate these functional groups with the various parts of the molecule.

4–12. Recent Developments

A. Mass Spectrometer/Computer Systems

The use of mass spectrometers on-line to computers is particularly attractive since mass spectra contain a lot of information which is readily digitalized (e.g., analogue electrical signal converted to digital m/e and intensity values).

(*i*) *Low resolution.* The signals from a low resolution scan can be fed directly into a computer. The computer memory can also be used to store the spectra of a selected range of organic compounds. The spectrum which has just been recorded can then automatically be compared with those in the memory, and the compound is often identified in this way. Another advantage is that low resolution digitalized spectra from the computer can be reproduced in the form of the bar graphs produced in this chapter using on-line automatic plotters.

(*ii*) *High resolution.* The double focusing mass spectrometer is used in the high resolution mode (narrow slits) and the signals fed into a computer which is programmed with the exact masses of C, H, O, N, etc. The time of arrival of the signals can automatically be converted to exact ion masses (± 2 ppm) and thence to ion compositions. Ion compositions, automatically printed out in order of increasing m/e values and sorted according to heteroatom content (e.g. separate columns for C, H, and C, H, O-containing ions), are known as element maps. Element maps contain a lot of information but (like many low resolution spectra) may be difficult to interpret in detail; the computer may be used to aid interpretation of the spectrum. One disadvantage of high resolution operation is a loss in sensitivity.

B. *Mass Spectrometer/Gas Chromatography Systems*

The effluent from a gas chromatograph is usually a pure compound (or, less frequently, a simple mixture) in small quantity (10^{-12}–10^{-3} g.) in the gas phase. The quantity and condition is ideal for direct transfer into a mass spectrometer source, providing the carrier gas (e.g., He) can first be largely removed. The removal of carrier gas is effected by the use of separators which rely on either (i) the faster diffusion of He (e.g., through a sinter) or (ii) preferential passage of the organic substrate through a membrane. A schematic illustration of a two-stage membrane separator is shown in Fig. 4–23. The gas chromatograph is thus directly coupled to the mass spectrometer, and the spectrum of each component is obtained as it leaves the column. Identification of substances in very small quantities (e.g. 10^{-9}–10^{-12} g.) is thus possible. The identification of steroids and drug metabolites in urine are important examples.

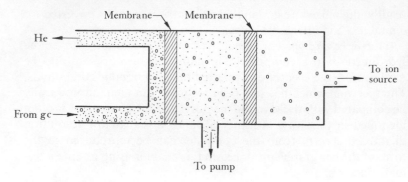

Fig. 4–23

C. Other Ionization Techniques

The following alternatives to ionization by electron impact are still being developed, but already seem to possess advantages in certain cases.

(*i*) *Field ionization and field desorption.* In field ionization the substance being studied, in the gas phase, is ionized by an electric field of the order of 10^7–10^8 volts/cm. Ionization occurs at an anode, which may be a sharp blade, sharp tip or thin wire. Source residence times are very short (*ca.* 10^{-12} sec.) and the internal energies of the molecular ions low. Therefore daughter ions are of low abundance and molecular ion peaks are intense.

If the molecule to be examined is involatile or thermally unstable (e.g. glucose), then the substance may be placed directly on the anode and 'thrown' into the gas phase as a positive molecular ion repelled by the anode. The technique is known as field desorption. It produces, for example, mainly M^+ and $M^+ + 1$ ions from glucose, even though no molecular ion is obtained with this compound upon electron impact.

(*ii*) *Chemical ionization.* The substance to be examined is present in the source at pressures *ca.* 10^{-4} torr, but a gas such as methane is present at much higher pressures (~ 1 torr). Electron impact causes mainly ionization of methane, which fragments in part to CH_3^+. These species then undergo the following ion–molecule reactions at the high source pressures used.

$$CH_4^+ + CH_4 \longrightarrow CH_5^+ + CH_3$$
$$CH_3^+ + CH_4 \longrightarrow C_2H_5^+ + H_2$$

CH_5^+ can then act as a Bronsted acid, and $C_2H_5^+$ as a Lewis acid to produce ions from the substrate. The substrate ions so produced usually have sufficient energy to fragment, and a chemical ionization mass spectrum is produced. One advantage of this technique is that basic nitrogen compounds protonate very readily and often give abundant $M^+ + 1$ ions (quasi-molecular ions) in their spectra, whereas M^1 ions may be of negligible abundance or absent in electron impact spectra, for example,

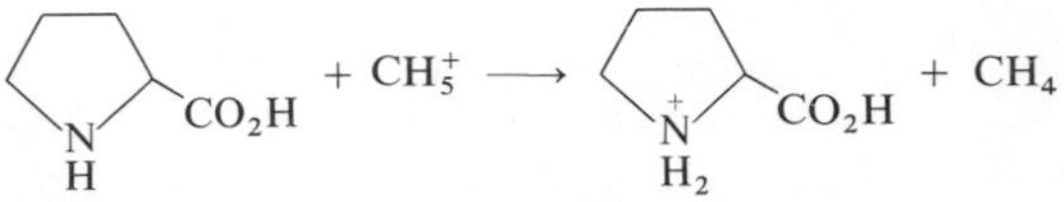

Quasi-molecular
ion ($M^+ + 1$)

D. Peptide Sequencing

Compounds which have molecular weights in the range 300–1000, and contain numerous polar functional groups such as OH, NH, SH, are normally derivatized, prior to examination in the mass spectrometer, so as to increase their volatility. Peptides provide a good example, and these are normally first acetylated (with acetic anhydride/methanol) and then permethylated (upon treatment with, say, dimethyl sulphoxide anion ($\overline{C}H_2SOCH_3$) followed by methyl iodide). Thus the peptide consisting of the amino acid sequence leucine.glutamic acid.glutamine.valine.proline.tyrosine would be derivatized as shown below and in particular this treatment increases the volatility by removing the zwitterionic nature of the peptide and removing intermolecular H-bonding ($\diagdown$NH----O=C$\diagup$). Upon electron impact, the modified peptides then break down from the molecular ion largely *via* cleavage of amide bonds with charge-retention by the *N*-terminus (to give *sequence ions*). Since the derivatized amino acids within the peptide have characteristic masses, the difference in mass between successive sequence ions allows the sequence of amino acids in the peptide to be determined (see Fig. 4–24).

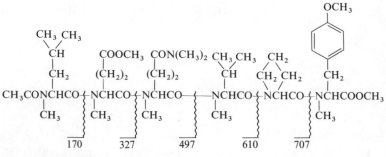

(i) Ac₂O/MeOH
(ii) ⁻CH₂SOCH₃/MeI

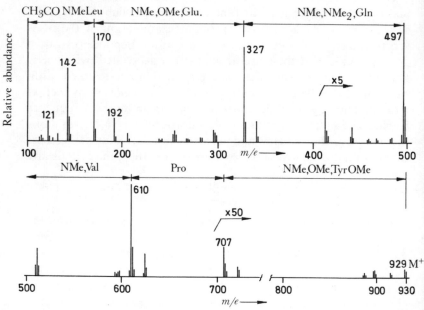

Fig. 4–24

The presence of tyrosine is also indicated by ions at m/e 121 and 192. The ion at m/e 142 arises via CO loss from the sequence ion at m/e 170.

$$CH_3O\!-\!\!\langle\rangle\!-\!\overset{+}{C}H_2 \qquad\qquad CH_3O\!-\!\!\langle\rangle\!-\!CH\!=\!CH\!-\!COOCH_3 \; \rceil^{\ddot{}}$$

m/e 121 $\qquad\qquad\qquad\qquad\qquad\qquad$ m/e 192

This approach is successful with the majority of peptides consisting of up to 8 amino acid residues. It can be applied at the 100 μg. level, and a particular advantage is that it can be applied to simple mixtures of peptides.

Bibliography

F. H. Field and J. L. Franklin, *Electron Impact Phenomena*, Academic Press, New York, 1957.

J. H. Beynon, *Mass Spectrometry and Its Applications to Organic Chemistry*, Elsevier, Amsterdam, 1960.

K. Biemann, *Mass Spectrometry*, McGraw-Hill, New York, 1962.

C. A. McDowell (ed.), *Mass Spectrometry*, McGraw-Hill, New York, 1963.

H. Budzikiewicz, C. Djerassi and D. H. Williams, *Structure Elucidation of Natural Products by Mass Spectrometry*, Vols. I and II, Holden-Day, San Francisco, 1964.

F. W. McLafferty, *Interpretation of Mass Spectra*, Benjamin, New York, 1966.

H. C. Hill, *Introduction to Mass Spectrometry*, Heyden and Son, 1966.

G. Spiteller, *Massenspectrometrische Structuranalyse Organischer Verbindungen*, Verlag Chemie, Weinheim, 1966.

H. Budzikiewicz, C. Djerassi and D. H. Williams, *Mass Spectra of Organic Compounds*, Holden-Day, San Francisco, 1967.

J. H. Beynon, R. A. Saunders and A. E. Williams, *The Mass Spectra of Organic Molecules*, Elsevier, London, 1968.

A. L. Burlingame (ed.), *Topics in Organic Mass Spectrometry*, Wiley-Interscience, New York, 1970.

Q. N. Porter and J. Baldas, *Mass Spectrometry of Heterocyclic Compounds*, Wiley-Interscience, London, 1971.

G. W. A. Milne (ed.), *Mass Spectrometry—Techniques and Applications*, Wiley-Interscience, New York, 1971.

G. R. Waller (ed.), *Biomedical Applications of Mass Spectrometry*, Wiley-Interscience, New York, 1972.

D. H. Williams and I. Howe, *Principles of Organic Mass Spectrometry*, McGraw-Hill, 1972.

Tables for Use in Mass Spectrometry

J. H. Beynon and A. E. Williams, *Mass and Abundance Tables for Use in Mass Spectrometry*, Elsevier, Amsterdam, 1963.

J. Lederberg, *Computation of Molecular Formulas for Mass Spectrometry*, Holden-Day, San Francisco, 1964.

J. H. Beynon, R. A. Saunders and A. E. Williams, *Table of Metastable Transitions for Use in Mass Spectrometry*, Elsevier, Amsterdam, 1965.

5. Structure Elucidation by Joint Application of UV, IR, NMR, and Mass Spectroscopy

5–1. General Approach. 5–2. Worked Examples. 5–3. Problems.

5–1. General Approach

A definite sequence may be followed in applying those spectro-scopic methods which have been discussed in detail in chapters 1–4 to the structure elucidation of organic compounds. In the first place, the molecular formula of the compound is known from exact mass measurements on the molecular ion in the mass spectrum; in the absence of a molecular ion, this information can be ascertained by micro-analysis. From the molecular formula, the number of *double bond equivalents* (D.B.E., that is the degree of unsaturation as contained in double bonds and/or rings) in the molecule may be calculated. Equation 5–1 holds for molecules containing C,H,(O) only, whereas equation 5–2 is applicable to molecules containing N.

For $C_aH_b(O_c)$: D.B.E. $= \dfrac{(2a+2)-b}{2}$ (5–1)

For $C_aH_b(O_c)N_d$: D.B.E. $= \dfrac{(2a+2)-(b-d)}{2}$ (5–2)

In both equations, the $(2a+2)$ factor represents the corresponding saturated hydrocarbon. For a C,H,(O) compound the number of double bond equivalents is then obtained by subtracting the actual number of hydrogen atoms present (b) from $(2a+2)$ and dividing by 2, as in equation 5–1. Since nitrogen is trivalent, a hydrogen atom is subtracted from b for each nitrogen atom in the molecular formula (equation 5–2). In applying equations 5–1 and 5–2, monovalent elements (e.g., Cl, Br, I) may be replaced by H; divalent elements may effectively be ignored; trivalent P is treated as N;

and for pentavalent P three hydrogen atoms must be subtracted from b.

The number of double bond equivalents calculated as indicated above may serve as a valuable guide in visualizing plausible structures from a molecular formula. For example, a molecular formula $C_{10}H_{11}N_3$ establishes the presence of seven double bond equivalents (often abbreviated as ⎡7⎤), to which highly unsaturated structures such as I and II correspond. It is useful to remember that a benzene ring is associated with four double bond equivalents (one ring and three double bonds).

I   II

Secondly, the UV spectrum may give some indication of the type of chromophore present. Thirdly, the IR spectrum may identify (or, for that matter, show the absence of) certain functional groups. Finally, the way in which these functional groups and the remaining atoms are joined together can be deduced from the mass spectrum and the NMR spectrum, with occasional support from the fingerprint region of the IR spectrum. The sequence of analysis is illustrated in the following pages with four worked examples. These examples are not the very easiest spectra to interpret; you should already, from reading Chapters 1–4, have learned how the simpler elements of an organic structure can be identified: you could try, for example, to predict what the four spectra of such compounds as diethyl ether, methyl ethyl ketone, isopropanol, and benzyl acetate would look like. In the examples which follow, we have chosen to show how powerful the methods are in identifying quickly and unambiguously the complete structures of four slightly more complex compounds. The spectra of twelve more compounds are then presented for you to try to deduce unambiguous structures from them, although in two cases you will find a small ambiguity that cannot be removed.

The NMR spectra for the examples and problems in this chapter have been obtained at 100 MHz. The scales are marked in units of δ (0–10 ppm); the small scale divisions are 0·1 ppm or 10 Hz, since

1 ppm corresponds to 100 Hz at 100 MHz. All the data accumulated in the tables at the end of chapter 3 (chemical shifts in δ units and coupling constants in Hz) are directly applicable to these spectra without any conversion factors. The spectra are not actually labelled IR, UV, NMR or Mass, since it is clear which is which.

5–2. Worked Examples

Example 1

The formula, $C_8H_8O_2$, obtained from the mass spectrum shows that five double bond equivalents are present. The presence of a highly unsaturated system is further confirmed from the ultraviolet spectrum which shows a strong band at 316 nm (ε 22,000). The band strength is calculated from equation 1–2 using the concentration and path length shown on the spectrum and the molecular weight:

$$\varepsilon = \frac{0 \cdot 43 \times 136 \times 100}{0 \cdot 265} = 22,000$$

The long wavelength absorption of this band and the intensity are consistent with about four or five conjugated double bonds.

The IR spectrum shows that there are few, if any, saturated C—H's (only weak bands at just below 3000 cm.$^{-1}$) and some aryl or unsaturated C—H's (the weak band at 3100 cm.$^{-1}$). Since this latter absorption is inherently weak (whereas the former is usually strong) it seems likely that most of the hydrogen atoms are bound to unsaturated centres. The carbonyl region shows two bands: one at 1695 cm.$^{-1}$ and one at 1675 cm.$^{-1}$. These are likely to be $\alpha\beta$-unsaturated or aryl ketone, aldehyde or acid groups. The last of these is eliminated by the absence of the H-bonded —OH absorption in the 3000–2500 cm.$^{-1}$ region, and an aldehyde group is ruled out by the absence of absorption in the $\delta = 9$–10 region of the NMR spectrum. The compound is, therefore, probably a ketone. The strength and position of the band at 1615 cm.$^{-1}$ shows that a C=C conjugated double bond or conjugated aryl group must be present, and bands at 1555 cm.$^{-1}$ (too weak to be a nitro group) and 1480 cm.$^{-1}$ also suggest an aromatic type of compound, though probably not in the form of a benzene ring. Since we know from both the UV and the molecular formula that double bonds are likely to be present, we may glance, tentatively, into the finger-print region near 960 cm.$^{-1}$, where a strong band (actually at 970 cm.$^{-1}$) is indeed present. This would support the presence of a

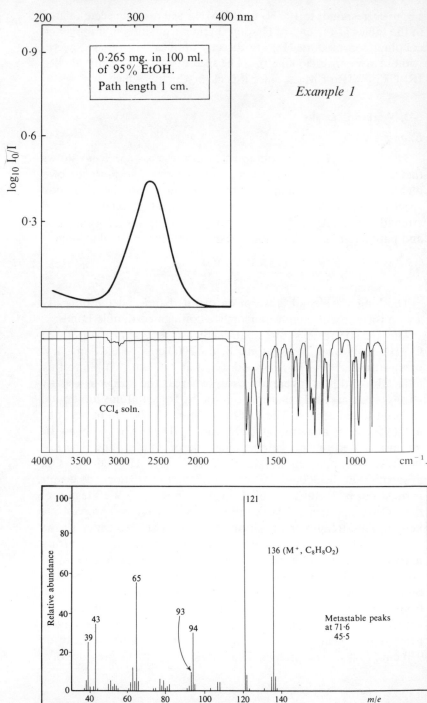

0·265 mg. in 100 ml. of 95% EtOH.

Path length 1 cm.

Example 1

$\log_{10} I_0/I$

200   300   400 nm

CCl_4 soln.

4000 3500 3000 2500 2000   1500   1000 cm^{-1}.

136 (M$^+$, $C_8H_8O_2$)

Metastable peaks
at 71·6
45·5

Relative abundance

m/e

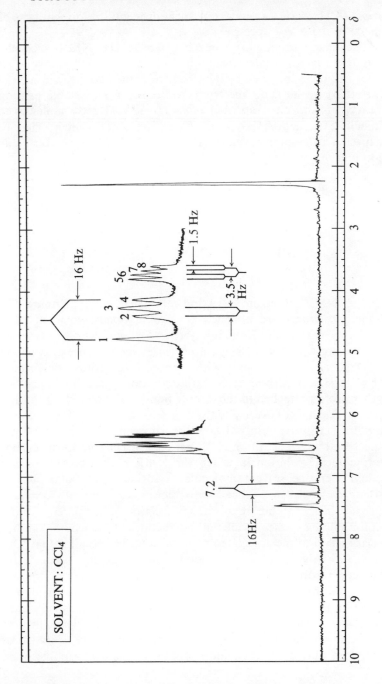

trans—CH=CH— system; but such an identification will best be made in the NMR spectrum.

The sharp singlet at $\delta = 2.29$ ppm in the NMR spectrum strongly suggests that the ketone is in fact a methyl ketone. This conclusion is supported by the presence of the base peak at $M-15$ (m/e 121) in the mass spectrum. Moreover, a metastable peak at 71·6 establishes the transition m/e 121$\rightarrow m/e$ 93 and enables us to infer the sequence III$\rightarrow a \rightarrow b$. A methyl ketone might also be expected to eliminate ketene (III$\rightarrow c$) and also to give rise to an ion d at m/e 43.

$$[C_6H_5O]\text{—}COCH_3 \longrightarrow [C_6H_5O]\text{—}C\equiv O^+ \xrightarrow[*]{-CO} [C_6H_5O]^+$$

$$\text{III} \qquad\qquad a,\ m/e\ 121 \qquad\qquad b,\ m/e\ 93$$

$$\Big\downarrow -CH_2=C=O$$

$$[C_6H_6O]^{\ddagger} \qquad\qquad\qquad CH_3C\equiv O^+$$

$$c,\ m/e\ 94 \qquad\qquad\qquad d,\ m/e\ 43$$

The NMR spectrum also indicates that the remaining five hydrogen atoms are attached to double bonds, some of them probably part of an aromatic system. The suggestion of a *trans*-double bond (from the IR) is confirmed by the presence of two lines (centred at $\delta = 7.2$ ppm) with a very large coupling constant ($J = 16$ Hz). In carbon tetrachloride solution, those lines due to the other proton attached to the double bond are obscured, since three proton resonances overlap in the $\delta = 6.3 - 6.7$ region. This complex pattern has been resolved (see the superimposed traces on the spectrum) by the addition of 10 per cent of benzene to the carbon tetrachloride solution in which the NMR spectrum was initially taken. Benzene solvent molecules associate with polar sites in solute molecules and hence locally they become partially oriented. Because benzene is highly anisotropic (section 3–4), different protons in the solute experience different magnetic fields depending on their relationship to the associated benzene ring. The result is that what were formerly complex overlapping signals may now separate and become resolved. If the superimposed trace on the left (normal sweep width) is expanded linearly by a factor of 4 (see superimposed trace on the right), eight lines due to the three protons are clearly resolved. Obviously, line 1 does not correspond to a whole proton; but it is separated by 16 Hz from line 4, and lines 1 and 4 are therefore due to the remaining proton on the *trans*-double bond, both protons making up an AB system. We have now detected systems IV and V (in which the groups X and Y must carry no protons which can couple vicinally with the protons of the *trans*-double bond).

IV   V   VI

The remainder of the molecule consists of C_4H_3O—, containing three double bond equivalents; two of these are likely to be double bonds (from the UV), while the third could, conceivably, be due to a ring or a double bond. The presence of a ring seems much more likely, since three double bonds carrying only three hydrogen atoms would have to be unreasonably highly substituted (to satisfy the molecular formula). The C_4H_3O group is therefore most plausibly a monosubstituted furan (VI). The chemical shift values of the three furan protons (6·46, 6·62 and 7·48 ppm) suggest an α-substituted furan (one proton in a more electronegative environment than the other two). The splittings of the three furan protons do in fact provide ample confirmation of structure VII; H_α is a poorly resolved doublet ($\delta = 7\cdot48$ ppm in CCl_4, $J_{\alpha\beta} = 1\cdot5$ Hz— see section 3–8A), $H_{\beta'}$ is also a doublet (lines 2 and 3, $J_{\beta\beta'} = 3\cdot5$ Hz), whereas H_β is a quartet (lines 5, 6, 7 and 8, $J_{\beta\beta'} = 3\cdot5$ Hz and $J_{\alpha\beta} = 1\cdot5$ Hz).

VII   VIII

The two carbonyl bands in the IR spectrum are probably due to the presence of the s-*trans*-(VII) and s-*cis*-(VIII) conformers. Interconversion between VII and VIII is rapid at room temperature and the NMR spectrum therefore presents only a time-averaged picture; it might be possible to see the spectra of both conformers

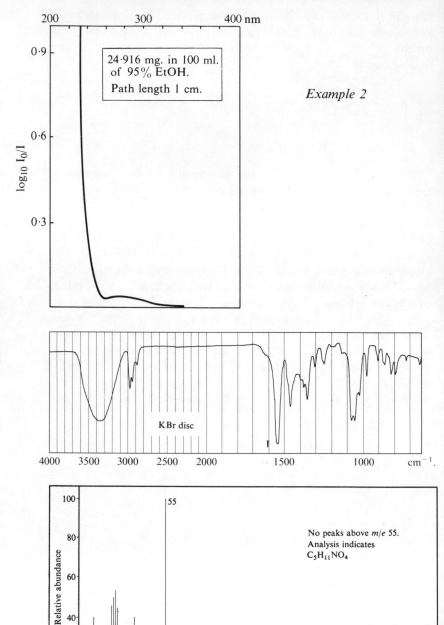

200   300   400 nm

0·9

24·916 mg. in 100 ml.
of 95% EtOH.
Path length 1 cm.

Example 2

0·6

$\log_{10} I_0/I$

0·3

KBr disc

4000   3500   3000   2500   2000   1500   1000   cm^{-1}.

100   55

80

60

Relative abundance

40

20

0
          20   40   60   m/e

No peaks above m/e 55.
Analysis indicates
$C_5H_{11}NO_4$

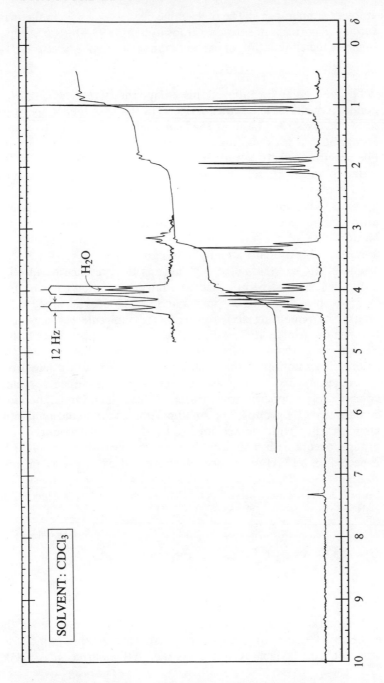

at lower temperatures. The metastable peak at m/e 45·5 in the mass spectrum corresponds to the transition m/e 93→m/e 65, which suggests the elimination of carbon monoxide from b to give $C_5H_5^+$.

Example 2

The molecular formula of this compound, $C_5H_{11}NO_4$, corresponds to one double bond equivalent. In this case the IR spectrum shows a very strong band at 1545 cm.$^{-1}$ due, very likely, to a nitro group. Such a formulation is supported, in the first place, by the UV spectrum where a very weak band (ε is calculated to be 24) is due to an n→π* transition (a ketone cannot be present since no band is evident in the 1800–1600 cm.$^{-1}$ region) and, secondly, by the mass spectrum, since a parent ion is usually absent with aliphatic nitro compounds. The nitro group therefore accounts for all the unsaturation of the molecule. The IR spectrum also shows the unmistakeable presence of a hydroxyl group.

The NMR spectrum strongly suggests the presence of an ethyl group, which must be attached to a fully substituted carbon atom (1:2:1 triplet at $\delta = 1·00$ ppm and 1:3:3:1 quartet at $\delta = 2·00$ ppm). The remaining six protons in the molecule resonate as a two-proton triplet at 3·32 ppm and as a four-proton eight-line pattern centred at 4·12 ppm (see integral trace). When the deuterochloroform solution of the compound is shaken with a few drops of deuterium oxide, the two-proton signal disappears almost completely (see superimposed trace). The compound must therefore contain two OH groups. The small portion of the signal remaining after shaking with deuterium oxide, due to unexchanged OH groups, occurs with a slightly different chemical shift, as might be expected for an OH resonance. Since the hydroxyl groups' protons resonate as a triplet, both alcohol groups must be primary. The units (IX–XI) which have been deduced can be fitted together in only one way to give the structure XII.

$$-NO_2 \qquad Y\!-\!\underset{\displaystyle Z}{\overset{\displaystyle X}{\underset{|}{\overset{|}{C}}}}\!-\!CH_2CH_3 \qquad -(CH_2OH)_2 \qquad HOCH_2\!-\!\underset{\displaystyle NO_2}{\overset{\displaystyle CH_2CH_3}{\underset{|}{\overset{|}{C}}}}\!-\!CH_2OH$$

| IX | X | XI | XII |

A remarkable feature of the NMR spectrum is that after the addition of deuterium oxide, the methylene protons which are adjacent to oxygen resonate as an AB quartet ($\delta_A = 4·00$,

$\delta_B = 4{\cdot}24$ ppm, $J_{AB} = 12$ Hz). The two protons of a methylene group adjacent to an asymmetrically substituted carbon atom (XIII) may thereby be magnetically non-equivalent, and spin–spin coupling may therefore be observed between them. This phenomenon is, moreover, not restricted to an asymmetric carbon atom, but may be shown by methylene protons adjacent to any dissymmetric moiety. Hence, the H_A and H_B protons of XIV are magnetically non-equivalent, because either of the $-CH_2OH$ groups 'sees' three *different* groups attached to the central carbon atom. In XIV, the methylene protons of the ethyl group are equivalent.

$$
\begin{array}{c}
\text{X} \\
| \\
\text{Y---C---CH}_2\text{---R} \\
| \\
\text{Z}
\end{array}
\qquad\qquad
\begin{array}{c}
\text{H}_A \quad\; \text{CH}_2\text{CH}_3 \;\; \text{H}_A \\
|\qquad\quad | \qquad\qquad | \\
\text{HO---C------C------------C---OH} \\
|\qquad\quad | \qquad\qquad | \\
\text{H}_B \quad\;\; \text{NO}_2 \qquad\;\; \text{H}_B
\end{array}
$$

$$\text{XIII} \qquad\qquad\qquad\qquad \text{XIV}$$

Additional multiplicity of the H_A and H_B resonances is observed before the deuteration experiment, because both H_A and H_B are coupled to the hydroxyl proton ($J = 6$ Hz). It should be remembered that the observation of coupling, in deuterochloroform or carbon tetrachloride solution, between the protons of a CH—OH group is the exception rather than the rule (see, for example, Fig. 2–4). The resonance at $\delta = 3{\cdot}99$ ppm, which appears in the superimposed trace, is due to traces of water present in the deuterium oxide. The peak at m/e 55 in the mass spectrum is due to $C_4H_7^+$.

Example 3

The molecular formula shows the presence of two double bond equivalents, and the absence of UV absorption shows that there are not two *conjugated* double bonds. The IR spectrum shows a strong carbonyl band at 1740 cm.$^{-1}$ which could be a five-ring ketone or a saturated ester; but the absence of the weak n→π* transition in the UV shows that the compound must be an ester. This is supported by the presence of strong bands in the C—O stretching region 1300–1100 cm.$^{-1}$. Since there is no C=C double bond absorption the two double bond equivalents in the compound can be accommodated by a diester, by a six-ring or larger lactone with two ether oxygens in the molecule (there is no —OH absorption and no other carbonyl absorption in the IR, leaving only the ether type of structure in which to dispose the oxygen atoms), or a monoester diether with another ring.

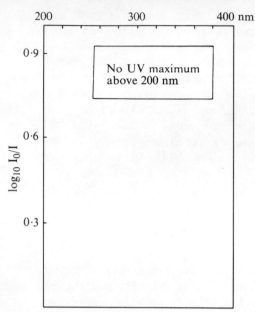

No UV maximum
above 200 nm

Example 3

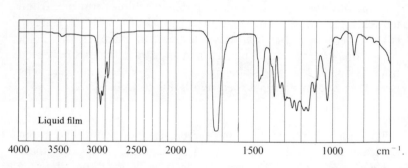

Liquid film

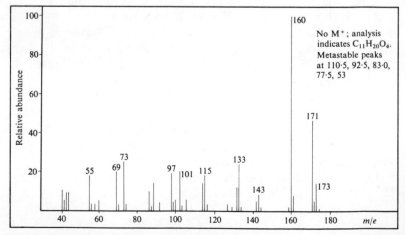

No M$^+$; analysis
indicates $C_{11}H_{20}O_4$.
Metastable peaks
at 110·5, 92·5, 83·0,
77·5, 53

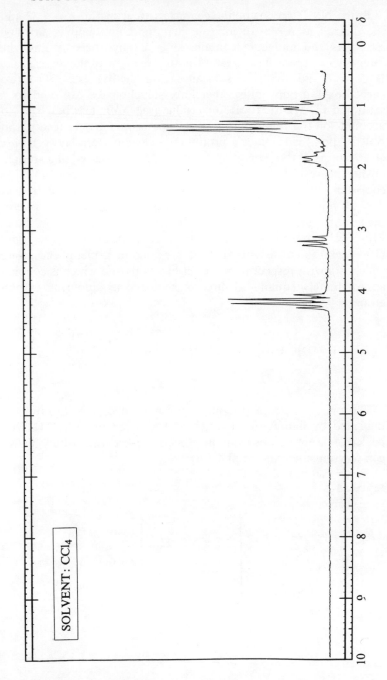

SOLVENT: CCl₄

In the NMR spectrum, the characteristic triplet at $\delta = 1{\cdot}32$ and the quartet at $\delta = 4{\cdot}16$ indicate that the compound is an ethyl ester. As the molecule contains only 20 hydrogen atoms, the intensities of these signals establish that the first of the alternatives is correct and that the substance is a diethyl ester. The two carbethoxyl groups must, therefore, somehow be attached to a saturated C_5H_{10}-hydrocarbon residue (see XV). The best way to account for the relatively low field one-proton triplet (centred at $\delta = 3{\cdot}2$ ppm) is to place a proton on a carbon atom between *both* ester groups. If the remaining carbon atoms are placed in a straight chain, the resulting structure [diethyl n-butylmalonate (XVI)] is consistent with the presence of the perturbed triplet centred at $\delta = 1{\cdot}0$ ppm with $J = 7$ Hz; this feature is quite characteristic of the terminal methyl group of a linear saturated hydrocarbon chain. Moreover, the two-proton quartet centred at $\delta = 1{\cdot}88$ ppm ($J = 7$ Hz) is consistent with the presence of a methylene group which is β with respect to two ester functions and which is coupled approximately equally to three protons on neighbouring carbon atoms.

$C_5H_{10}(CO_2C_2H_5)_2$

XV

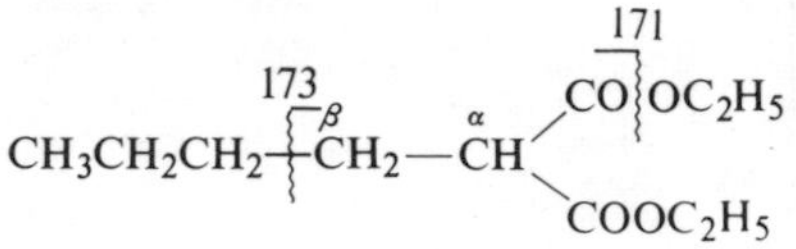

171
173
β
α
CO OC₂H₅
CH₃CH₂CH₂ CH₂—CH
COOC₂H₅

XVI

The mass spectrum provides confirmation of the structure, as indicated by figures in structure XVI and scheme 5–1. The base peak (*e*, *m*/*e* 160) arises from the process of β-cleavage with γ-hydrogen rearrangement (section 4–8B).

Scheme 5–1

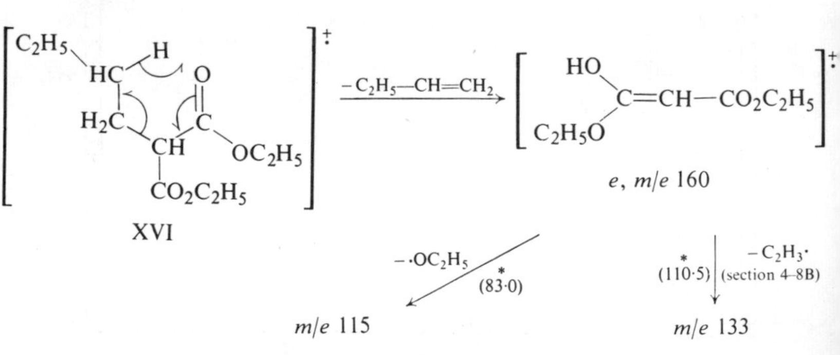

C₂H₅
HC
H
O
H₂C
C
CH
OC₂H₅
CO₂C₂H₅
XVI
−C₂H₅—CH=CH₂
HO
C=CH—CO₂C₂H₅
C₂H₅O
e, m/e 160
−·OC₂H₅
(83·0)
m/e 115
(110·5)
−C₂H₃·
(section 4–8B)
m/e 133

Example 4

This compound, $C_8H_{15}NO_2$, has two double bond equivalents, which are easily identifiable from the IR spectrum as a carbonyl group (1725 cm.$^{-1}$) and a $C=C$ double bond (1645 cm.$^{-1}$). Although the UV spectrum shows no maximum on the instrument used, it does show strong end absorption and must, therefore, have the $C=C$ double bond and the $C=O$ groups in conjugation; the ε value at 220 nm works out as 12,600, showing that the maximum must be shortly below 220 nm, since this is the order of magnitude for the ε value of two double bonds conjugated together. If, then, the carbonyl group is conjugated, the value of the IR band shows that it must be an $\alpha\beta$-unsaturated ester. A feature of this IR spectrum is the strong bands at 2770 and 2820 cm.$^{-1}$, which is the region in which —OMe, $>$NMe, —OCH$_2$— and $>$NCH$_2$—groups absorb. The strength of this band relative to the usual stretching band at 2950 cm.$^{-1}$ suggests a structure in which such groupings predominate.

In the mass spectrum, a very large percentage of the total ion current is carried by one ion at m/e 58 ($C_3H_8N^+$ from exact mass measurements) which, on account of the very favourable α-cleavage in basic nitrogen compounds (Table 4–6), almost certainly corresponds to f or g. The ion f could only arise in one manner (XVII→f), but ion g may be derived in two ways (XVIII→g or XIX→h→g).

XVII f, m/e 58

XVIII g, m/e 58

XIX h g, m/e 58

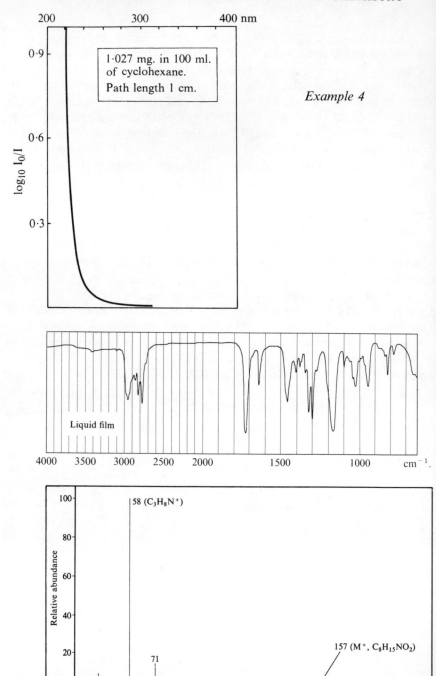

1·027 mg. in 100 ml. of cyclohexane.
Path length 1 cm.

Example 4

Liquid film

58 ($C_3H_8N^+$)

71

157 (M^+, $C_8H_{15}NO_2$)

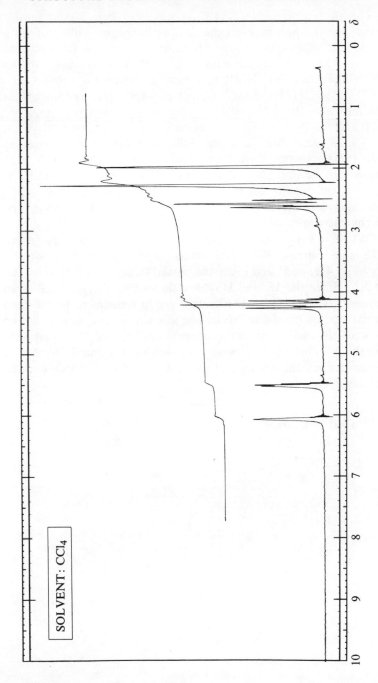

SOLVENT: CCl₄

The partial structure XVIII is ruled out because no proton resonance disappears when the sample is shaken with deuterium oxide and because there is no NH absorption in the IR spectrum. The integral indicates the number of protons associated with the six resonances in the NMR spectrum, as follows: $\delta = 6\cdot09$ (1 H); $5\cdot52$ (1 H); $4\cdot20$ (2 H); $2\cdot58$ (2 H); $2\cdot28$ and $1\cdot98$ (9 H). The integral in the region of the $2\cdot28$ and $1\cdot98$ resonances cannot be interpreted very accurately, but since this region corresponds to nine protons, it is likely that there are two equivalent methyl resonances at $2\cdot28$ (6 H) and one methyl resonance at $1\cdot98$ (3 H). No N-ethyl group can be present, and structure XIX is untenable (XVIII is again ruled out on this basis). The six-proton singlet at $\delta = 2\cdot28$ ppm therefore corresponds to the $-N(CH_3)_2$ group of XVII. The methylene group adjacent to nitrogen resonates at $2\cdot58$ ppm as a triplet (J = 6Hz), and is consequently adjacent to a CH_2 group. The same splitting (J = 6 Hz) occurs for the downfield two-proton triplet at $4\cdot20$ ppm, and hence the partial structure can be extended to XX. Since the IR and UV have uncovered the presence of an $\alpha\beta$-unsaturated ester, only a methyl group remains to be attached to the double bond (see XXI). The absence of any large coupling between the olefinic proton resonances (at $5\cdot52$ and $6\cdot09$ ppm) rules out the possibilities of a *cis*- or *trans*-substituted double bond, and hence defines the compound as N,N-dimethylaminoethyl methacrylate (XXII).

$$(CH_3)_2NCH_2CH_2O-(C_4H_5O)$$
$$\text{XX}$$

$$(CH_3)_2NCH_2CH_2OC\overset{\displaystyle O}{\overset{\|}{}}-C\overset{\displaystyle H}{\underset{}{}}\!\!\!\!=CH$$
$$\text{XXI}$$

$$(CH_3)_2NCH_2CH_2OC\overset{\displaystyle O}{\overset{\|}{}}-C\overset{\displaystyle CH_3}{\underset{}{}}=CH_2$$
$$\text{XXII}$$

5–3. Problems

In the following twelve problems the conditions under which the spectra were obtained are indicated on the actual spectra. Unless otherwise stated, no changes were observed in the NMR spectra after the solutions had been shaken with deuterium oxide. All metastable peaks that were observed are quoted; some of these, however, may not be useful in interpretation using only the information given in chapter 4.

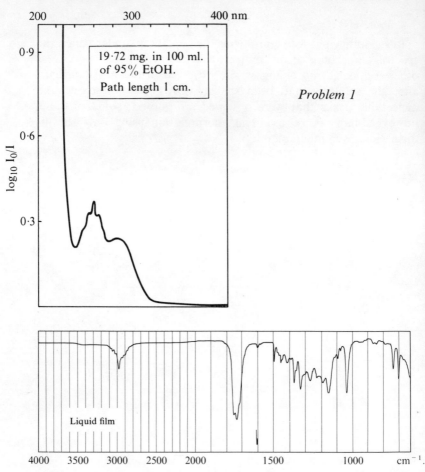

19·72 mg. in 100 ml.
of 95% EtOH.

Path length 1 cm.

Problem 1

Liquid film

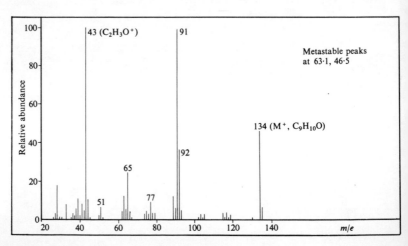

43 ($C_2H_3O^+$)

91

Metastable peaks
at 63·1, 46·5

134 (M^+, $C_9H_{10}O$)

92

65

51

77

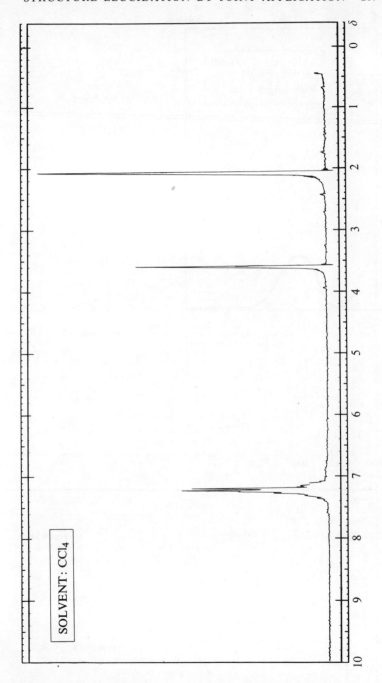

SOLVENT: CCl$_4$

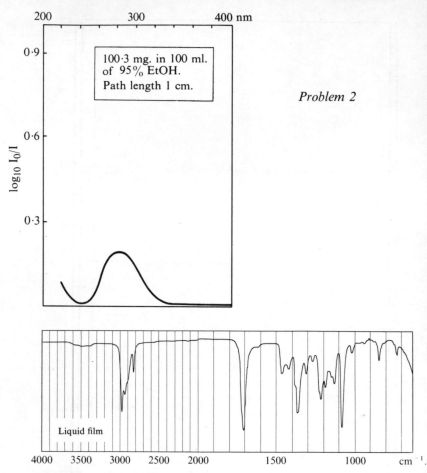

200   300   400 nm

$\log_{10} I_0/I$

0.9

0.6

0.3

100·3 mg. in 100 ml.
of 95% EtOH.
Path length 1 cm.

Problem 2

4000   3500   3000   2500   2000   1500   1000   cm^{-1}.

Liquid film

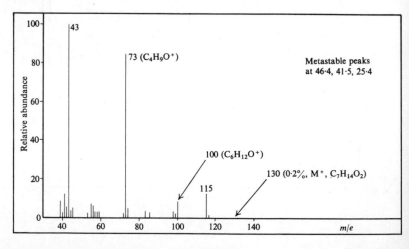

43

73 (C$_4$H$_9$O$^+$)

Metastable peaks
at 46·4, 41·5, 25·4

100 (C$_6$H$_{12}$O$^+$)

130 (0·2%, M$^+$, C$_7$H$_{14}$O$_2$)

115

Relative abundance

100

80

60

40

20

0

40   60   80   100   120   140 m/e

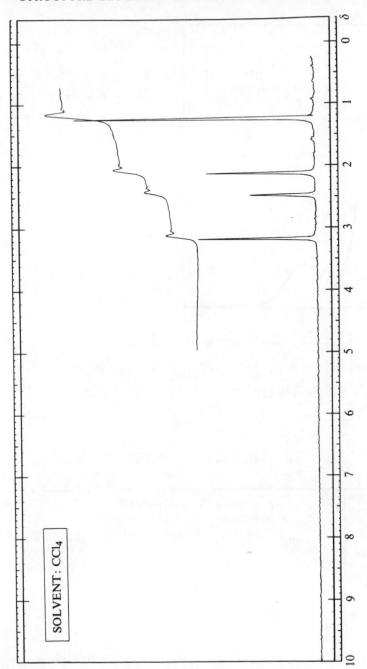

SOLVENT: CCl₄

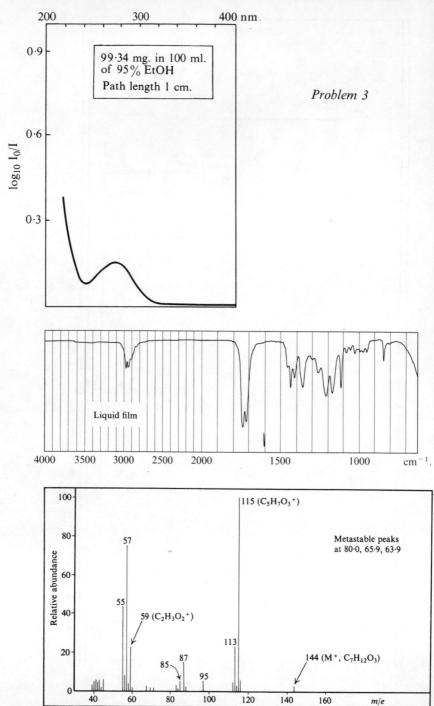

99·34 mg. in 100 ml.
of 95% EtOH
Path length 1 cm.

Problem 3

Liquid film

115 ($C_5H_7O_3^+$)

Metastable peaks
at 80·0, 65·9, 63·9

57

55

59 ($C_2H_3O_2^+$)

85   87

95

113

144 (M^+, $C_7H_{12}O_3$)

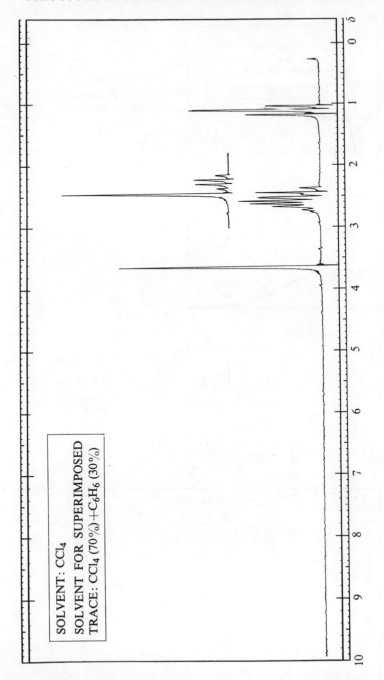

SOLVENT: CCl_4
SOLVENT FOR SUPERIMPOSED
TRACE: CCl_4 (70%) + C_6H_6 (30%)

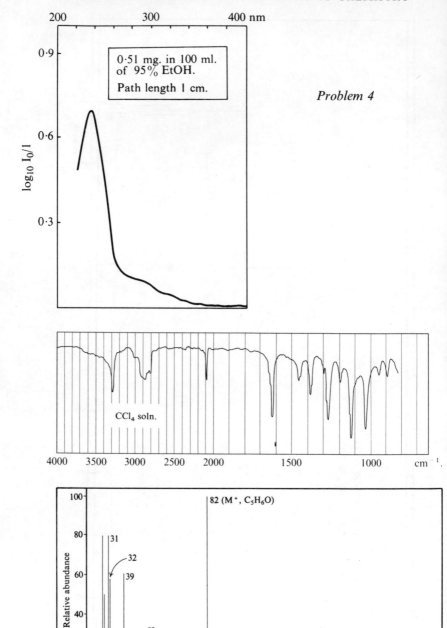

0·51 mg. in 100 ml. of 95% EtOH.

Path length 1 cm.

Problem 4

CCl_4 soln.

82 (M^+, C_5H_6O)

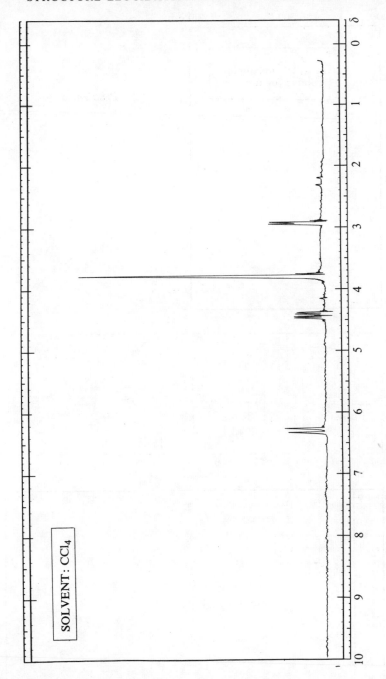

SOLVENT : CCl₄

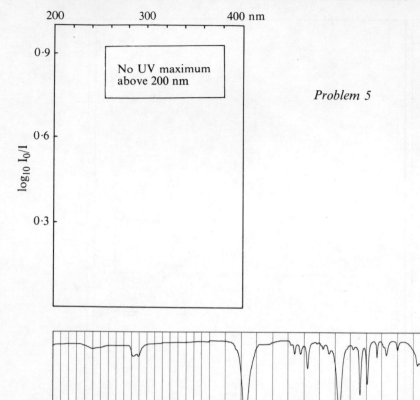

No UV maximum above 200 nm

Problem 5

$\log_{10} I_0/I$

Liquid film

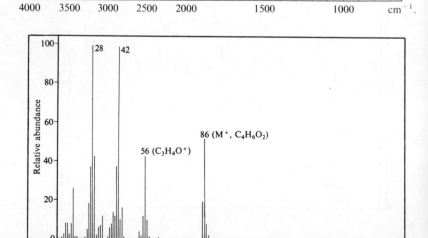

Relative abundance

28
42
56 ($C_3H_4O^+$)
86 (M^+, $C_4H_6O_2$)

m/e

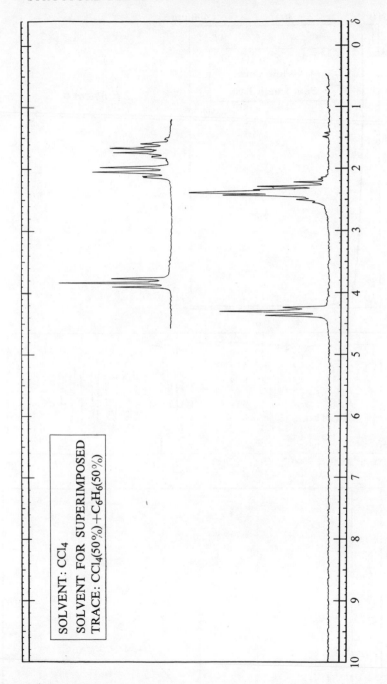

SOLVENT: CCl$_4$
SOLVENT FOR SUPERIMPOSED
TRACE: CCl$_4$(50%)+C$_6$H$_6$(50%)

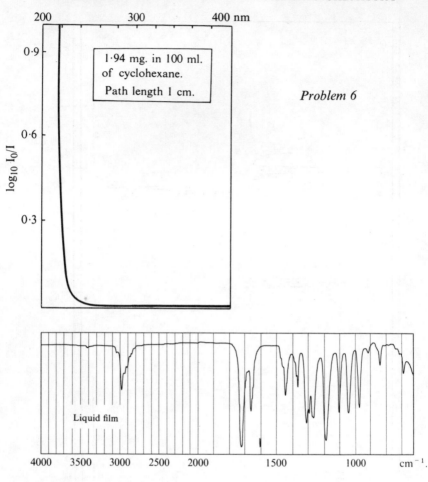

1·94 mg. in 100 ml. of cyclohexane.
Path length 1 cm.

Problem 6

Liquid film

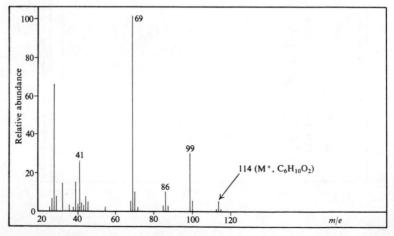

114 (M$^+$, C$_6$H$_{10}$O$_2$)

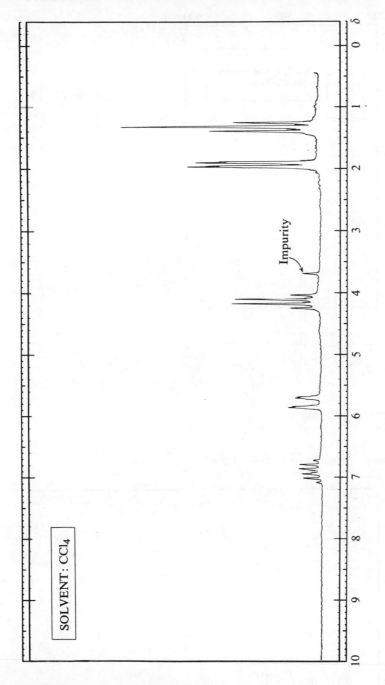

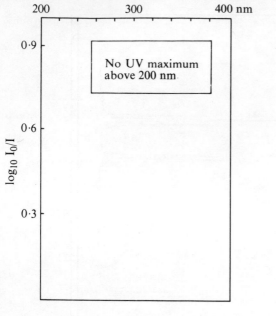

No UV maximum above 200 nm

Problem 7

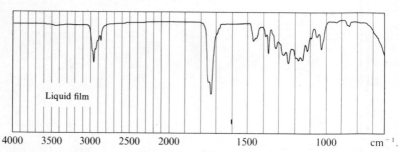

Liquid film

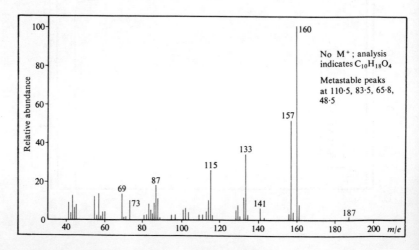

No M$^+$; analysis indicates $C_{10}H_{18}O_4$

Metastable peaks at 110·5, 83·5, 65·8, 48·5

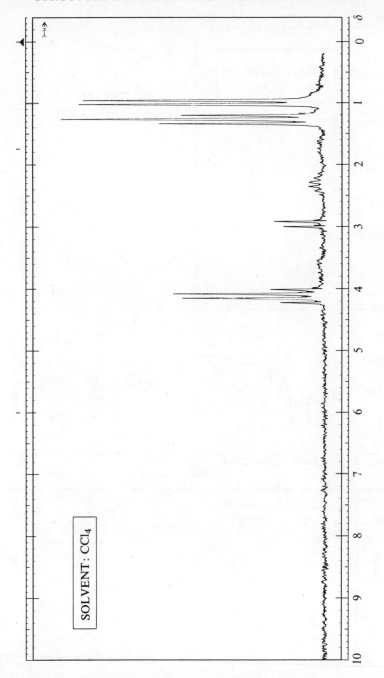

SOLVENT: CCl$_4$

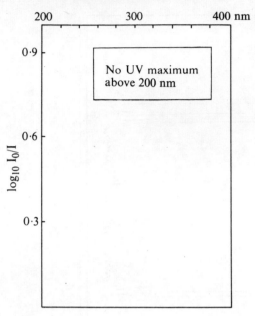

No UV maximum
above 200 nm

Problem 8

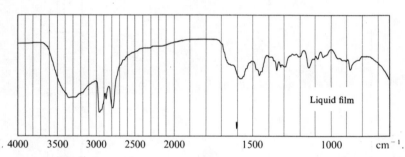

Liquid film

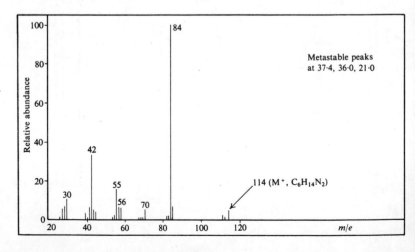

Metastable peaks
at 37·4, 36·0, 21·0

114 (M$^+$, $C_6H_{14}N_2$)

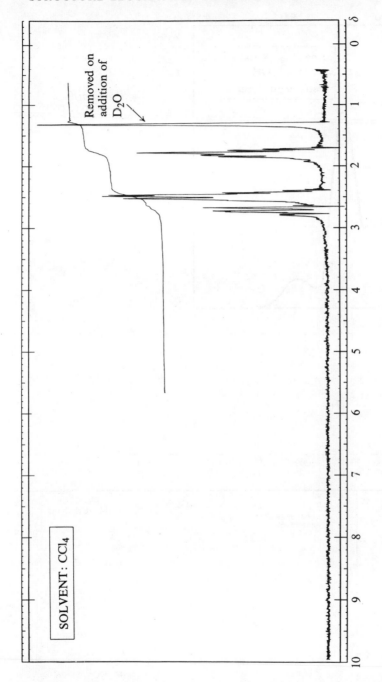

SOLVENT: CCl₄

Removed on addition of D₂O

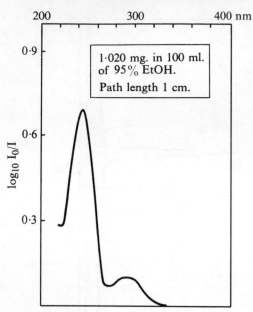

1·020 mg. in 100 ml.
of 95% EtOH.

Path length 1 cm.

Problem 9

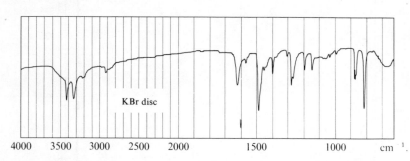

KBr disc

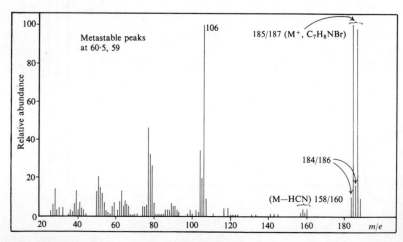

Metastable peaks
at 60·5, 59

185/187 (M$^+$, C_7H_8NBr)

184/186

(M−HCN) 158/160

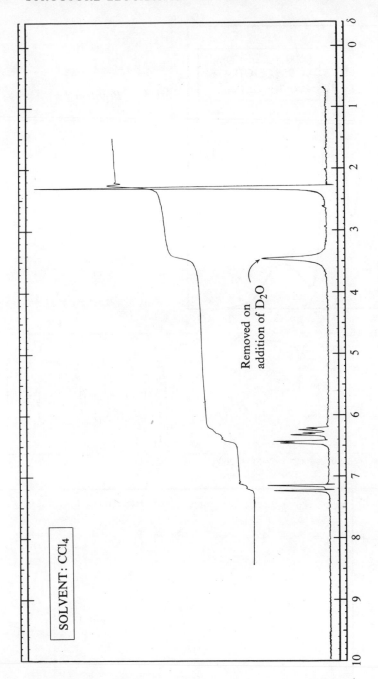

SOLVENT : CCl₄

Removed on addition of D₂O

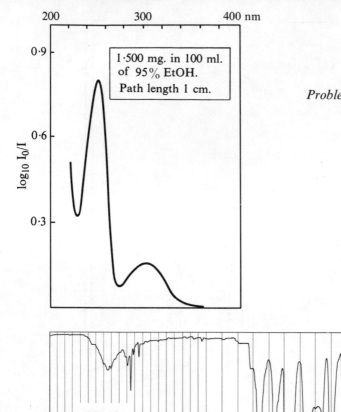

1·500 mg. in 100 ml. of 95% EtOH. Path length 1 cm.

Problem 10

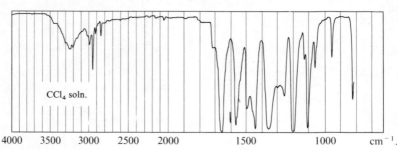

CCl₄ soln.

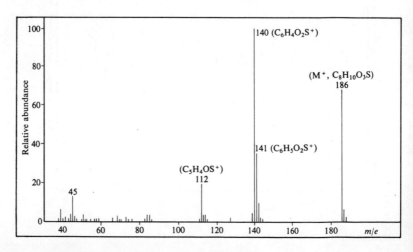

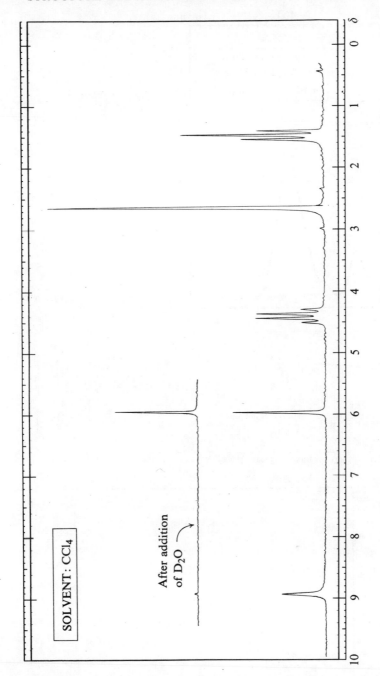

SOLVENT: CCl₄

After addition
of D₂O

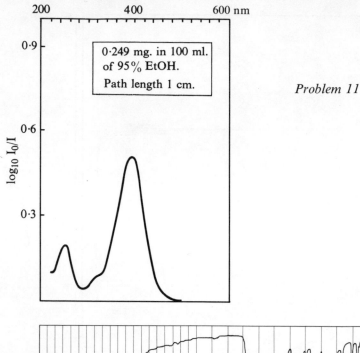

0·249 mg. in 100 ml.
of 95% EtOH.
Path length 1 cm.

Problem 11

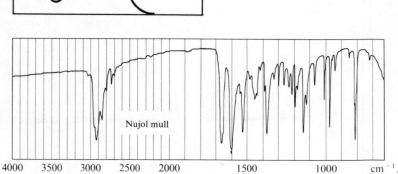

Nujol mull

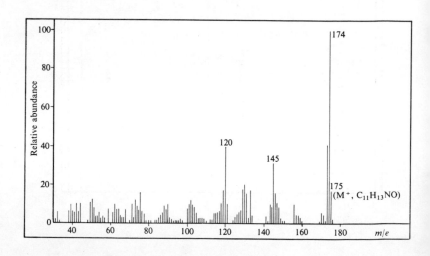

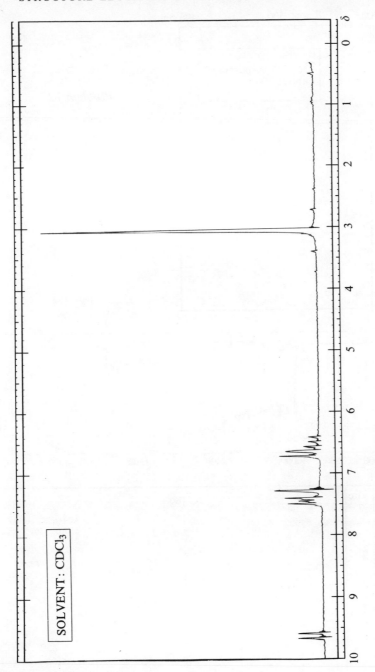

SOLVENT: CDCl₃

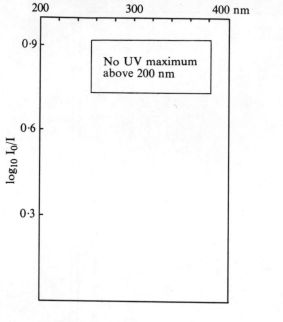

No UV maximum above 200 nm

Problem 12

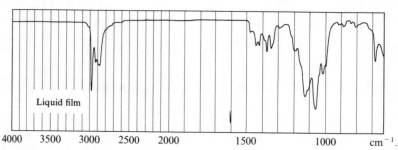

Liquid film

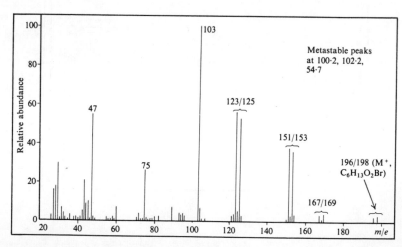

Metastable peaks at 100·2, 102·2, 54·7

196/198 (M$^+$, $C_6H_{13}O_2Br$)

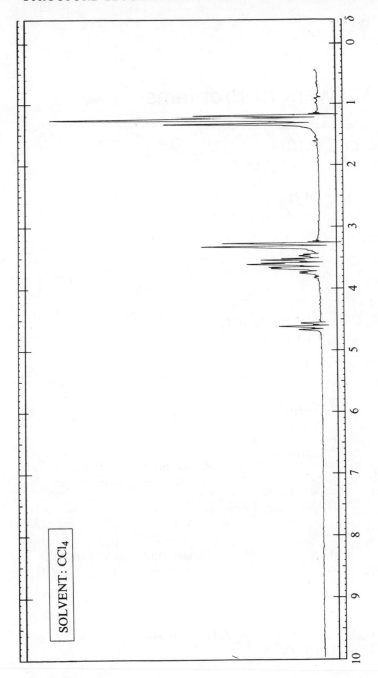

SOLVENT: CCl$_4$

Answers to Problems

1. CH_3COCH_2Ph

2. $(CH_3)_2C\text{—}CH_2COCH_3$
$$\qquad\quad |$$
$$\qquad\; OCH_3$$

3. $C_2H_5COCH_2CH_2CO_2CH_3$

4. $CH_3OCH\overset{c}{=\!=}CH\text{—}C\!\equiv\!CH$

5.

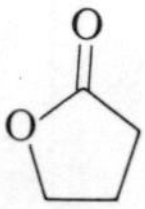

6. $CH_3CH\overset{t}{=\!=}CHCO_2C_2H_5$

7. $(CH_3)_2CHCH(CO_2C_2H_5)_2$

8.

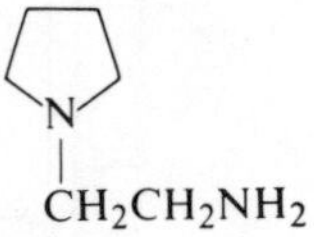

$\overset{|}{CH_2CH_2NH_2}$

9.

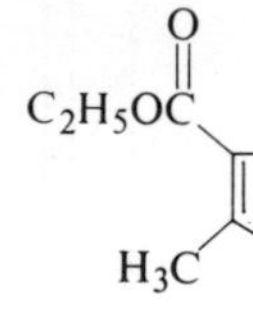

Other bromotoluidines with a 1,2,4-substitution pattern are possible.

10.

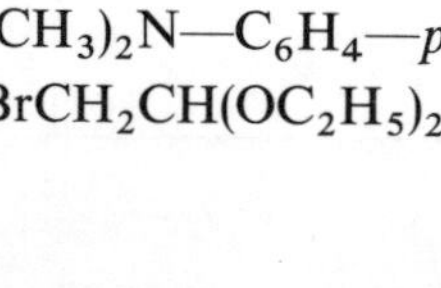

Other thiophenes with the carbethoxyl group adjacent to CH_3 and/or OH are possible.

11. $(CH_3)_2N\text{—}C_6H_4\text{—}p\text{—}CH\overset{t}{=\!=}CHCHO$

12. $BrCH_2CH(OC_2H_5)_2$

Index

Printed in Great Britain by
William Clowes & Sons, Limited
London, Beccles and Colchester